# CyberTruck

*Owner's manual*

# Contents

1. General Misinformation .................................................................. 8
    Learn how to tell apart a Cybertruck from a large metallic toaster. ............ 9
    Debunking myths like "Is it a spaceship?" ..................................... 12
    Dubious Vehicle Specifications .................................................. 14
    The Power of Exaggeration ....................................................... 17
    Size Matters, Or Does It ........................................................ 19
2. Exterior Illusions and Interior Delusions ...................................... 22
    It's Not a Tank, It's Just Ugly ................................................ 25
    The Inside Story: Navigating the labyrinth of an interior .................... 28
    Instruments and Guesswork ....................................................... 31
    Confusing Control Panel Arrangement ........................................... 34
    The Maze of Buttons ............................................................. 37
    Knobs and Dials Galore .......................................................... 40
    The Single Button of Uncertain Functions ....................................... 42
    Will it launch the car or just the windscreen wipers? ......................... 44
    A game of guessing what the button does each time you press it. ............. 46
    The "What Does This Do?" Holographic Display ................................. 48
    Futuristic Confusion ............................................................. 51
    Trying to figure out if it's a map, a movie, or an abstract art piece. ...... 54
3. Starting and Stopping ............................................................. 57
    The Guessing Game of Engine Ignition ........................................... 60
    Is today the day it decides to start, or will it just make whirring noises? .62
    The Key to Nowhere ............................................................... 65
    Asphalt Ambiguity vs. Off-Road Roulette ........................................ 67

    Deciding whether "Sport" mode actually means "Slightly Less Slow." ...... 69

    Taking bets on whether "Off-Road" mode is actually just "Panic" mode. 71

    Autopilot: Trusting Your Life to Software .................................................. 73

    A Leap of Faith ............................................................................................. 75

    The Art of Letting Go .................................................................................. 77

4. Discomfort Systems ...................................................................................... 80

    Unpredictable Climate Control Shenanigans ............................................ 83

    The Sauna vs. Arctic Tundra Setting .......................................................... 85

    Guess which vent will actually blow air today! ......................................... 87

    Seat "Adjustments" via Esoteric Levers .................................................... 89

    Pull a lever and see what moves - if anything. ......................................... 91

    Comfort? More Like a Torture Device ....................................................... 93

    Mood Lighting That Reflects Your Disappointment ................................. 94

    Discover the lighting that barely illuminates but maximizes despair ........ 96

    When the lights decide to have a rave without your consent. ................ 98

5. Entertainment or Lack Thereof ................................................................... 101

    The AM Radio Experience: Static and All ................................................ 103

    How to find that one station that's not just white noise. ....................... 105

    Embracing the static as part of your musical journey. ............................ 107

    Pairing Your Device with the Infuriotainment System ............................ 109

    Bluetooth or Blue-rage? ............................................................................ 111

    The Eternal "Loading" Screen: Patience is a virtue ................................. 113

    Navigating the Abyss of the User-Unfriendly Interface ........................... 115

    A Maze of Menus: Delve into the labyrinthine interface ........................ 118

    The Mystery of the Missing Features ...................................................... 120

6. "Routine" Maintenance ............................................................. 122
    Realigning the Wheels: A Guide to Guesswork ........................... 125
    The Alignment Lottery ................................................................ 127
    When DIY Stands for Destroy It Yourself ................................... 129
    Battery Maintenance: A Shocking Surprise ............................... 131
    Is your battery charging, or is it just pretending? .................... 133
    Jump-Start Jamboree ................................................................ 135
    Software "Upgrades": Crossing Your Fingers and Toes ............ 137
    Update or Downgrade ............................................................... 139
    The Eternal Reboot ................................................................... 141

7. Emergency Pretenses ............................................................. 144
    How to Utilize the Decorative Parachute .................................. 147
    Understanding that the parachute is more of a concept than a reality..148
    False Hopes and Dreams ........................................................... 150
    Deploying the Theoretical Safety Bubble ................................. 153
    Discover the safety features that exist only in This manual......155
    Is it a safety bubble or just a bubble of hot air? ...................... 157
    The Ejector Seat: For Exiting Conversations About Mileage ... 159
    Learning the art of ejecting yourself from tedious talks about fuel efficiency. ................................................................................. 161
    Use at Your Own Risk ............................................................... 163

8. Driver "Aids" and Assumptions .............................................. 166
    Parking Assistance: The Bumper's Sacrifice ............................ 169
    The Crunching Sound of Assistance ......................................... 171

The Art of Guesswork: How to park using your instincts because the sensors are probably wrong. ................................................................... 173

Hill Start: A Game of Chance .............................................................. 175

Rolling the Dice: Will you move forward, roll back, or just stay put? ..... 178

The Thrill of Uncertainty .................................................................... 180

The Invisibility Cloak: Just Close Your Eyes ......................................... 182

Pretending your car can become invisible, because why not? ............. 184

Exploring the exciting possibility of not seeing things right in front of you. ................................................................................................... 186

9. Vehicle Neglect ................................................................................ 188

The Myth of Scratch Removal ............................................................ 191

Scratches: A Badge of Honor .............................................................. 193

The Futility of Buffing: Why trying to remove scratches might just give you more. ......................................................................................... 196

The Eternal Quest for a Clean Windscreen ........................................ 199

Mastering the art of achieving a streak-free windscreen, or at least trying to. ..................................................................................................... 201

The ongoing battle between your wipers and the elements. ............. 203

Anti-Rust Coating: Also Known As Mud .............................................. 205

Understanding how a layer of mud can be just as effective as any anti-rust coating. ..................................................................................... 207

Embrace the Dirt ................................................................................ 209

10. Dubious Upgrades and Custom Jobs ............................................. 211

Adding a Spoiler for Absolutely No Reason ........................................ 214

The Illusion of Speed: Convincing yourself that a spoiler makes your car go faster. ............................................................................................ 216

    Style Over Substance ............................................................................ 219

    Customizing the Horn to Be Even More Annoying ............................... 222

    The Symphony of Irritation ................................................................. 224

    Horn Etiquette, or Lack Thereof .......................................................... 226

    The Ill-advised Nitrous Oxide Button ................................................. 228

    More Fizz Than Fury .......................................................................... 230

    The False Rocket Launch .................................................................. 232

11. Technical Mysteries ................................................................................ 234

    Interpreting Warning Lights as Modern Art ........................................ 237

    The Dashboard Disco .......................................................................... 240

    Warning Lights or Christmas Decorations .......................................... 242

    The Random Reset: Automotive Roulette ........................................... 244

    Discovering the joys of your car's systems resetting at the most inopportune moments. ........................................................................ 246

    Reset and Hope for the Best ................................................................ 248

    When to Simply Give Up and Call a Tow Truck ................................ 250

    The Inevitable Surrender ..................................................................... 252

    Tow Truck Tales .................................................................................. 255

12. Warranty Weaseling ................................................................................ 257

    Limited Warranty: Very Limited Indeed ............................................. 259

    Discovering the many, many exclusions of your "comprehensive" warranty. ............................................................................................. 262

    Warranty or Lottery ............................................................................. 264

    Service Schedules: More Like Guidelines .......................................... 267

    Understanding that service schedules are more suggestions than actual rules...........................................................................................................269

    Learning why putting off service can be a risky game. ...........................271

    Customer "Support": An Exercise in Patience..........................................273

    Finding inner peace while listening to the same tune for hours on end.275

    Scripted Responses and Frustration..........................................................278

13. Addendum of Afterthoughts ..................................................................281

    List of Unusable Tools Included with Your Vehicle ..................................283

    A comprehensive guide to tools that are for display purposes only........285

    Understanding why the supplied jack doesn't actually fit your car. ........287

    Incomprehensible Software Licensing ......................................................289

    Decoding the End User License Agreement, one headache at a time. ...291

    Why your car's software thinks it's still living in the 1990s.....................292

    How to Lodge a Complaint Into the Void .................................................294

    The Art of Yelling at Clouds ......................................................................296

    Crafting the Perfect Unread Email ...........................................................297

# 1. General Misinformation

Learn how to tell apart a Cybertruck from a large metallic toaster.

Welcome, brave soul, to the curious world of the Cybertruck, Tesla's latest escapade into making vehicles that double as conversation pieces at pretentious cocktail parties. Now, you might be thinking, "Why on Earth do I need to learn how to differentiate between a Cybertruck and a toaster?"

Well, let's just say, if you can't tell your breakfast appliance from your vehicle, you might be in for a bit of a shock – and not the electric kind.

First and foremost, size matters. A toaster, unless you're living in a world where bread is baked in slabs the size of a small country, is usually small enough to fit snugly on your kitchen counter. The Cybertruck, on the other hand, is about as subtle as a hippo in a tutu. It's big, it's bulky, and it looks like it was designed by a child who just discovered the wonders of a ruler. If it looks like it could flatten your neighbor's Prius, it's probably not a toaster.

Now, let's talk about aesthetics. The Cybertruck looks like it was designed for a dystopian future where the only entertainment is watching robots play dodgeball. It's all sharp angles and flat planes, like a stealth bomber had a love affair with an IKEA flat-pack. A toaster, by comparison, is usually a bit more rounded, a bit more welcoming. It's like comparing a cactus to a pillow; one is decidedly more comfortable to embrace.

Then there's the matter of functionality. A toaster, in its mundane brilliance, turns bread into toast. That's pretty much it. It's a one-trick pony, but what a trick it is. The Cybertruck, meanwhile, promises to haul, tow, and generally act like the burly truck it wants to be when it grows up. If you can fit your vehicle in a kitchen cabinet, or if it has a setting for bagels, it's probably not a Cybertruck.

But let's not forget the windows. Ah, the windows. Tesla claimed they were unbreakable, which was quickly proven wrong in a most public and rather embarrassing fashion. If you throw a metal ball at your toaster and it shatters, congratulations, it's not a Cybertruck – but you might need a new toaster. And possibly a lesson in anger management.

Of course, there's the small matter of electricity. Yes, both items run on it, but plug a Cybertruck into your kitchen socket and you'll likely be greeted by a spectacular light show, followed by darkness and the smell of fried wiring. The toaster, on the other hand, is quite content with the humble offerings of your kitchen's electrical system. Remember, just because they both use electrons, doesn't mean they're interchangeable.

In terms of interior, the Cybertruck is rather spartan, like a monk's cell with a steering wheel. It's all very functional and futuristic, in a 'I-gave-up-comfort-for-style' kind of way. Your toaster, unless you're living in a particularly avant-garde household, doesn't come with seating options or a touchscreen. If you find yourself sitting in your toaster, you may have other, more pressing issues to address.

The driving experience is also somewhat different. If you're behind the wheel of a Cybertruck, you'll likely feel a sense of power, a hint of smugness, and the sneaking suspicion that you're in a sci-fi movie. Driving a toaster, however, is generally frowned upon and highly impractical, unless it's part of an art installation or a particularly vivid dream.

Lastly, the price tag. The Cybertruck, while cheaper than some of Tesla's other offerings, still requires a bank account that doesn't wince at the sight of a comma in the price. A toaster, on the other hand, can often be acquired for the price of a decent meal. If you're haggling over the price and the seller is using words like 'financing' and 'down payment', you're probably not in the toaster aisle.

In conclusion, while both the Cybertruck and a large metallic toaster might share some superficial similarities – like their propensity for electric current and their strikingly modern aesthetic – they are, in fact, as different as chalk and cheese, or perhaps more fittingly, as different as toast and trucks. Keep this guide in mind, and you'll never embarrassingly try to drive your toaster to work or butter your bread with the Cybertruck. At least, one would hope.

## Debunking myths like "Is it a spaceship?"

Welcome to the section of the Cybertruck manual where we delve into the wild, wacky, and outright ludicrous myths surrounding this hulking mass of electric bravado. You've heard the whispers, the rumors, the barroom theories – and now, it's time to squash them like a grape under a steamroller.

First up, "Is the Cybertruck a spaceship?" No, it is not. I know, I know, it looks like something Captain Kirk would drive to a Star Trek convention, but trust me, it's as much a spaceship as I am a ballerina. It doesn't fly, it doesn't hover, and it most certainly doesn't travel at warp speed. It has wheels. Four of them. And they stay very much on the ground, thank you very much. If you're expecting to escape Earth's gravity in a blaze of electric glory, prepare to be disappointed.

Now, onto the next: "Has the Cybertruck come from the future?" Again, no. If it had, it would come equipped with a manual written in some futuristic language like binary or emoji. It's a truck, not a time machine. It's made from steel, not unobtanium, and it was designed here on Earth by people who probably wear sneakers and drink overpriced coffee. It's about as futuristic as a microwave oven was in the 1980s – cutting edge for now, but give it a few years.

"But it looks like nothing else on the road!" you cry. True, but so did the AMC Pacer, and nobody's accusing that of being a relic from the year 3000. The Cybertruck's design is bold, brash, and breaks all the rules, but so does wearing a kilt to a funeral. It doesn't mean it's from another time period, it just means it's different.

Then there's the favorite of conspiracy theorists: "The Cybertruck is bulletproof, so it must be built for the apocalypse." Let me be clear, just because it can take a hit or two doesn't mean it's Mad Max's vehicle of choice. The truth is far less exciting. It's just a bit tougher than your average truck, designed for the kind of person who finds themselves in a hailstorm of golf balls. It's not an apocalyptic war machine; it's a precaution against wayward shopping carts and the occasional rock thrown by a disgruntled badger.

And what about the myth that the Cybertruck can float? Yes, it's electric, but no, it's not a boat. If you drive it into a lake, it will sink, and you'll look like a right plum. Electric does not equal amphibious. It's a truck, not a duck. It does many things, but paddling across the English Channel isn't one of them.

"But it's so big and heavy, it must be slow!" There's another myth for the bin. It's like saying Usain Bolt can't be fast because he's taller than most sprinters. The Cybertruck defies its size and weight with the kind of acceleration that would make a cheetah spit out its breakfast in disbelief. It's quick, despite looking like it should be parked in a quarry.

And then, my personal favorite: "The Cybertruck is so tough, it doesn't need maintenance." Ha! That's like saying I don't need to go to the dentist because I brush my teeth. Everything needs maintenance, especially something that's part battering ram, part electric miracle. It might not need oil changes, but it will need tire rotations, software updates, and every so often, a good talking to when it inevitably decides to act up.

In conclusion, the Cybertruck is not a spaceship, it hasn't time-traveled here, and it's not an armored vehicle for the end of the world. It's a truck, an electric one, with more edges than a geometry textbook and enough torque to make a strongman weep. It's a marvel, a mystery, and a bit of a madman, but it's as terrestrial as tea and as contemporary as your last Instagram post.

So, let's put these myths to bed, alongside the notion that driving a Cybertruck will make you instantly attractive – spoiler alert, it won't.

## Dubious Vehicle Specifications

If you've ever gazed upon the Cybertruck and thought, "That looks like the lovechild of a DeLorean and a garden shed," then you're not alone. But let's delve deeper into this rolling enigma's specifications, which are as dubious as the likelihood of finding a decent cup of tea in America.

First off, there's the size. The Cybertruck is massive. It's like Tesla saw a regular truck and said, "Yes, but can we make it in a size that terrifies small children and animals?" Parking this beast is akin to docking the International Space Station. You'll need a parking space the size of Wales and the patience of a saint.

Then we have the exterior. Made from stainless steel, it's allegedly bulletproof, which is useful if you're expecting to drive through a war zone on your way to buy milk. However, it's worth noting that while it might stop a bullet, it's not so great with a metal ball at a product launch. It's like building a tank out of origami; impressive on paper but less so in practice.

Speaking of the exterior, let's talk about those angles. I haven't seen this many sharp edges since I tried to assemble a bookshelf without reading the instructions. It's all very dramatic and makes you feel like you're driving a piece of modern art. The downside is that birds have nowhere to perch, and you'll slice your hand open every time you go to open the door.

Now, onto the performance. The Cybertruck claims to go from 0 to 60 mph faster than you can say, "Is that really necessary?" It's like strapping a rocket to a cathedral and lighting the fuse. The acceleration is so brisk, it'll rearrange your facial features and have your passengers questioning their life choices.

The towing capacity is another thing Tesla boasts about. This truck can supposedly tow an amount that's akin to lugging a small moon. It's all well and good until you actually need to tow something, and you realize you don't own anything heavy enough to test this claim unless you start dragging your house behind you.

As for the interior, it's as minimalist as a monk's living quarters. There are so few buttons, you might think they forgot to install them. It's all very clean and futuristic, but when you want to adjust the air conditioning, you'll need a degree in computer science and the patience of a Buddhist monk.

The range is impressive, but remember, it's based on ideal conditions, like driving on a flat road with a tailwind, inside a vacuum, on a sunny day. In real-world conditions, like a slight hill or a stiff breeze, expect the range to be as optimistic as my chances of winning a marathon.

And then there's the Autopilot feature. It sounds fancy, but it's less "fully autonomous flying car" and more "slightly attentive driving assistant." Relying on the Cybertruck's Autopilot is like asking a squirrel to do your taxes; it's technically possible, but you'll be on edge the entire time.

Let's not forget the off-road capabilities. The Cybertruck can apparently traverse Mars, which is great if you live on Mars. On Earth, it'll handle a dirt road, but so can a determined cyclist. The suspension is adjustable, which is like saying I'm adjustable; I can sit, stand, or lie down, but that doesn't make me an athlete.

Lastly, the price. It's surprisingly affordable for an electric leviathan, but remember, that's just the starting price. By the time you've added all the extras, like windows that don't shatter when sneezed at, you'll need a second mortgage.

In conclusion, the Cybertruck's specifications are as dubious as my ability to cook a gourmet meal. It's big, it's brash, it's impractical, and it's like nothing else on the road. But let's face it, if you're buying a Cybertruck, you're not doing it because it makes sense. You're doing it because you want to drive something that looks like it was designed on another planet. And in that regard, it delivers spectacularly.

## The Power of Exaggeration

In the fantastical world of automotive marketing, 'horsepower' is the magic word, the spell that turns ordinary men and women into salivating beasts craving speed and power. Now, let's talk about the Cybertruck's alleged 500 horsepower. Is it really 500 horses under the hood, or just five horses and a vividly optimistic dream?

First off, let's understand what horsepower is. It's a unit of measurement that's about as easy to grasp as a greased pig. Essentially, one horsepower is supposedly the power one horse can deliver. So, by that logic, 500 horsepower would be akin to having the equivalent of a medium-sized cavalry regiment at your disposal. But anyone who's ever seen a horse, let alone 500 of them, will know that this comparison is as ludicrous as expecting a pig to fly.

Tesla claims that the Cybertruck has the pulling power of, well, 500 horses. That sounds impressive until you realize that horsepower in the electric world is like calories in fast food – conveniently inflated. In reality, the Cybertruck's power is more akin to a spirited game of charades – lots of enthusiastic gestures but not quite the real thing.

Let's break it down. When you press the accelerator, what you're really doing is sending a polite request to the electric motors. These motors then consider your request, check with the battery, and finally decide just how much of those 500 horses they're going to let out of the stable. It's like having a team of horses, but most of the time, a good chunk of them are off having a tea break.

And about those electric motors – they're the silent heroes, tirelessly spinning away. But to say they equate to 500 actual horses is like saying my blender has the same power as a team of huskies. It's technically power, yes, but not the sort you could use to, say, plow a field or charge into battle.

Then there's the issue of torque. Torque in an electric vehicle is like the plot in a soap opera – it's there, and it's potent, but it doesn't make a lick of sense. The Cybertruck boasts about its torque, but what it doesn't tell you is that this torque is as elusive as a politician's promise. It comes in a flash and is gone before you can say, "Was that it?"

Also, consider the weight. The Cybertruck is heavy, about as heavy as a small moon or a particularly large hippo. All this horsepower has to lug around the automotive equivalent of an obese elephant. It's like asking those five horses to not only run the race but to carry the entire grandstand while they're at it.

And let's not forget the battery. The battery in the Cybertruck is like that one friend who promises to help you move house but suddenly remembers a dentist appointment that day. The battery promises all this power, but the moment you start having fun, it begins to get cold feet, worrying about range and longevity.

Now, consider how you'll actually use this power. Unless you're regularly participating in illegal street races or fleeing from bank heists, the truth is, you'll rarely tap into that mythical 500 horsepower. Most of the time, you'll be pottering about, using the equivalent of one very bored horse and a disinterested donkey.

And when it comes to charging this beast, those 500 horses become more like 500 snails. You'll plug it in, and the power trickles in at a pace that would make a glacier say, "Hurry up." It's like filling a swimming pool with a teaspoon – technically possible, but maddeningly impractical.

In the end, while the notion of 500 horsepower sounds like a stampede of wild mustangs, the reality is more akin to a leisurely trot. It's not about the raw power; it's about the illusion of it. It's the automotive equivalent of wearing a superhero costume to a job interview – it looks impressive, but you're not fooling anyone.

The Cybertruck's horsepower is like a magic trick – it dazzles and awes, but don't look too closely, or you'll see it's all mirrors and smoke. But then again, who buys a Cybertruck for practicality? You buy it for the dream, the fantasy, the idea that beneath your right foot lies a herd of wild horses, ready to gallop into the sunset. Just remember, those horses are more likely to be taking a nap than charging into the horizon.

## Size Matters, Or Does It

Let's address the elephant in the room, or should I say, the yacht in the garage? The Tesla Cybertruck, a vehicle so large it could have its own postcode. The question is, is it really as big as a small yacht, or is this just another case of automotive hyperbole?

From the moment you lay eyes on the Cybertruck, you're hit by its sheer size. It's like Tesla took a regular pickup truck and fed it a diet of steroids and protein shakes. Parking this behemoth is akin to berthing the Titanic in a paddling pool. You can forget about squeezing into those compact city spaces unless you fancy remodeling the cars parked either side.

But let's get into the specifics. The Cybertruck measures in at just over 19 feet long. To put that into perspective, that's about half the length of a bowling alley. You could host a small party in its shadow. It's so long, by the time you've walked from the front to the back, you might need to take a break and admire the scenery.

Width-wise, the Cybertruck is about as broad as the average ego at a Silicon Valley cocktail party. You'll be doing more than just dominating the road;

you'll be annexing it. Changing lanes in this thing isn't so much a maneuver as it is a territorial conquest.

Height is where things get really interesting. It's tall enough that, should you drive under a low bridge, you might just give it a new sunroof. It's like Tesla decided that what the world really needed was a pickup truck that could double as a mobile watchtower.

Now, about that yacht comparison. A small yacht is typically around 30 feet long. So, in terms of length, the Cybertruck isn't quite yacht territory, but it's certainly encroaching on small naval vessel. However, in terms of height and width, it's more akin to a yacht's lifeboat — if the lifeboat was designed by someone who only had a ruler and a set of squares for tools.

Inside the Cybertruck, the space is equally generous. There's enough room to swing several cats, not that I'd recommend it. The rear seats have enough legroom for a basketball team, and the front feels like you're piloting a spaceship — a spaceship designed by someone who thought luxury meant 'lots of metal and screens'.

The bed of the truck is another marvel. It's so spacious, you could probably rent it out on Airbnb. It's not just a truck bed; it's a dance floor, a storage unit, a mobile picnic area. You could carry a small forest's worth of lumber or, if you're feeling particularly adventurous, enough equipment for a small circus.

But here's the kicker — with great size comes great responsibility, mainly to not terrify other road users. Driving the Cybertruck requires the spatial awareness of a fighter pilot and the nerves of a bomb disposal expert. You'll need to constantly remind yourself that you're not actually driving a small yacht, even though it feels like it.

Maneuvering this leviathan in city traffic is like threading a needle while wearing boxing gloves. You'll find yourself making three-point turns more often than a confused tourist. And let's not even talk about parallel parking.

Attempting to parallel park the Cybertruck is like trying to do a ballet in clogs – technically possible, but nobody wants to see that.

In conclusion, while the Cybertruck may not be quite as large as a small yacht, it's certainly pushing the boundaries of what can be considered a reasonable size for a vehicle. It's huge, it's imposing, and it's about as subtle as a firework display in a library. So, is size everything? In the case of the Cybertruck, size isn't just everything; it's the only thing. Remember, when driving this titan of the tarmac, you're not just a driver; you're the captain of your very own land yacht. Just don't try to sail it.

## 2. Exterior Illusions and Interior Delusions

When Tesla unveiled the Cybertruck, the world paused, not in awe, but in collective bewilderment. Here we have a vehicle that looks like it was designed using a ruler and a protractor, and with about as much regard for aesthetics as a concrete block. This is the Exterior Illusion – a design so bold, it makes Marmite look universally palatable.

Let's start with the 'armored glass'. The Cybertruck's windows are supposed to be impervious to everything short of a direct hit from a howitzer. However, at the grand unveiling, a simple throw of a metal ball shattered these claims, quite literally. It's like claiming you're bulletproof but ducking when someone throws a croissant at you.

The body of the Cybertruck is made from cold-rolled stainless steel, which is great if you're planning on driving through a hail of gunfire, but not so much if you're just popping down to the shops. It's like wearing a suit of armor to a dinner party – impressive, but entirely unnecessary, and you'll struggle to sit down.

Now, the shape of the Cybertruck is something else. It's all straight lines and sharp angles. It's as if the designers were inspired by a child's drawing of a car, or perhaps a doorstop. This design leads to what I call 'the origami effect' – it looks interesting, but you're not entirely sure it's supposed to be a vehicle.

And then there's the size. The Cybertruck is big. Absurdly big. It's like driving a small bungalow around. You half expect to see a family of four and a Labrador living in the back. Its sheer scale creates an optical illusion; it looks like it's moving slowly, even when it's not, much like a distant airliner.

Moving on to the Interior Delusions. The inside of the Cybertruck is a stark contrast to its exterior. Where the outside is all harsh lines and industrial design, the interior is minimalist to the point of being spartan. It's like walking into a modern art gallery, only to find it's been robbed.

The dashboard is a marvel of simplicity, dominated by a single, giant touchscreen, which controls virtually everything. It's the vehicular equivalent of having a smartphone with only one app. The rest of the dashboard looks like it's been made from leftover marble from a billionaire's kitchen. It's cold, it's hard, and it's about as welcoming as a tax audit.

The seats are another story. They're comfortable enough, but they have the aesthetic appeal of a waiting room in a dentist's office. And there's so much space inside, you could hold a small yoga class. It's like Tesla couldn't decide whether they were designing a car or a studio apartment.

The steering wheel, if we can call it that, looks like it was borrowed from a spaceship. It's a rectangle. A rectangle. It's as if someone looked at a traditional steering wheel and thought, 'Yes, but how can we make this more awkward?'

Then there's the rearview mirror. In a normal car, this is a simple, functional item. In the Cybertruck, it's been replaced by a digital display because, apparently, using a mirror was just too straightforward. It's like using a drone to pass the salt.

And let's not forget the visibility, or lack thereof. The windows are narrow, like firing slits in a medieval castle. This creates a driving experience akin to navigating a tank through a medieval village. You feel safe, but you're not entirely sure what's going on outside.

In summary, the Cybertruck is a study in contradictions. Its exterior is a fortress on wheels, designed to withstand an apocalypse, while its interior is a Zen garden, designed to calm and soothe. It's like a bouncer with a degree in philosophy. The Cybertruck isn't just a vehicle; it's an experience, a statement, a rolling enigma. It's the automotive equivalent of a mullet: business in the front, party in the back, and confusion all around.

## It's Not a Tank, It's Just Ugly

When you first lay eyes on the Cybertruck, you might be forgiven for thinking that it was designed during a particularly aggressive game of Pictionary. This is Tesla's unique design philosophy, where brutalism meets confusion in a whirlwind romance. The Cybertruck doesn't just challenge traditional design norms; it takes those norms, ties them to a rocket, and fires them into the sun.

Let's start with the obvious: the Cybertruck looks like a tank that has been on a diet. Its angular, metallic body gives off a vibe that's part military bunker, part futuristic origami project. It's as if someone tried to design a vehicle using only a ruler, and then ran out of ruler. The result is a truck that looks like it's perpetually scowling at the world, as if it's just found out it wasn't invited to the cool cars' party.

The design of the Cybertruck throws caution and aerodynamics to the wind, much like a child throws spaghetti at a wall. It's a blatant disregard for the conventions of automotive design, which typically favors curves and lines that suggest movement and grace. The Cybertruck, in contrast, suggests that movement and grace were invited to the party, but they couldn't find anywhere to park.

The front of the Cybertruck is flat and featureless, like the expression on a bored bodyguard. It's so featureless, in fact, that you wonder if the designers simply forgot to finish it. It's a face only a mother could love, if the mother was a sheet of industrial-grade steel.

Moving to the sides, the Cybertruck continues its theme of 'brutalism meets confusion'. The sharp angles and straight lines are so severe, you could probably use the truck as a geometry teaching aid. It's as if the vehicle is daring you to try and lean casually against it, only to walk away with a collection of acute-angle-shaped bruises.

The rear of the Cybertruck is no less confusing. The sloping rear makes it look like the truck is perpetually trying to back away from its own design. It's as if the back of the truck is embarrassed by the front and is slowly trying to leave the scene.

Now, let's talk about the color, or rather, the lack of it. The Cybertruck comes in any color you want, as long as it's metallic gray. It's like they started to think about paint options and then decided that colors were too mainstream. The result is a vehicle that looks like it's been forged in the heart of a dying star, which is impressive if you're a black hole, less so if you're a pickup truck.

The wheels of the Cybertruck look like they've been stolen from a set of a sci-fi movie where the director was told to 'make it look futuristic, but within budget'. They're big, they're chunky, and they look like they could crush a Prius just by glancing at it.

Inside, the Cybertruck's design philosophy of 'brutalism meets confusion' continues unabated. The interior is so minimalist, it makes a monk's cell look cozy. The dashboard is a barren wasteland of space, interrupted only by a touchscreen that's so large, it might have its own weather system. It's like the designers were told to make the interior 'futuristic' and their only reference was a sci-fi movie from the 1980s.

The seats are utilitarian to a fault. It's as if comfort was a secondary concern to making sure the seats could be hosed down after a particularly messy apocalypse. They're not so much seats as they are slabs of material that have begrudgingly agreed to be sat upon.

The steering wheel, or as I like to call it, the 'steering rectangle', continues the theme. It's as if the designers looked at a traditional steering wheel and thought, 'How can we make this less intuitive and more likely to cause confusion and mild panic?'

In summary, the Cybertruck's design is a bold statement in 'brutalism meets confusion'. It's a vehicle that looks like it was designed for a dystopian future where the only design brief was 'make it look tough, and make sure no one can lean against it comfortably'. It's not a tank, it's just ugly. But it's a special kind of ugly – the kind that's so confident in its own skin, it becomes almost, dare I say, beautiful. Almost.

## The Inside Story: Navigating the labyrinth of an interior

When you first clamber into the cavernous expanse of the Cybertruck's interior, you could be forgiven for thinking you've stumbled into an M.C. Escher sketch. There's a sense of disorientation, as if the laws of physics and ergonomics had a mild disagreement. The interior of the Cybertruck doesn't just challenge conventional design; it takes conventional design outside and gives it a good thrashing.

The dashboard, a vast, unadorned slab of material, looms before you like the monolith from "2001: A Space Odyssey". It's so minimalistic, it makes a Zen garden look positively cluttered. You half expect to find a hidden compartment containing the meaning of life, but no, it's just a glovebox, and a rather small one at that.

Then there's the infamous touchscreen, the centerpiece of this futuristic tableau. It's enormous, glaring at you with the cold, unblinking eye of HAL 9000. This screen, your gateway to controlling virtually every aspect of the vehicle, is as user-friendly as a crossword puzzle in Ancient Greek. Adjusting the air conditioning requires a series of swipes and taps that would baffle even a teenager.

The seats, while visually appealing in a Spartan sort of way, have all the ergonomic comfort of a park bench designed by a committee that couldn't agree on what a bench is for. They're angular, firm, and apparently upholstered with a material that was chosen for its ability to survive a nuclear blast rather than for anything as pedestrian as comfort.

Looking up, you'll find the rearview mirror has been replaced by a digital display, because why use a simple mirror when you can have a screen? It's like replacing your bathroom mirror with a tablet; it's impressive technology, but you can't help but feel it's solving a problem that didn't exist.

The steering wheel, or rather, the steering rectangle, defies all conventional wisdom. It suggests that somewhere, in a design meeting, someone said, "You know what's wrong with steering wheels? They're just too wheel-like." It's a bold design choice, and one that makes you feel like you're driving a spaceship rather than a car – a spaceship designed by someone who's never actually driven anything.

As for the window controls and door handles, they're hidden away with the kind of zeal usually reserved for state secrets. Finding them is like playing a game of hide and seek with a particularly petulant child. You know they're there somewhere, but they're not going to make it easy for you.

The pedals are where you expect them to be, which is a relief, but they're about the only thing that meets your expectations. The accelerator feels less like a pedal and more like a suggestion box. You press it and hope the car takes your request under advisement.

The cup holders, an essential feature in any vehicle, appear to have been added as an afterthought. They're awkwardly placed, as if the designers begrudgingly acknowledged that some people might – heaven forbid – want to drink something while driving.

And then there's the space. So much space. The Cybertruck's interior is like a vast, empty warehouse. There's enough room to swing a cat, a dog, and a small pony. You could hold a small dance party in there, or start a new hobby, like indoor archery.

The storage compartments are equally perplexing. They're either cavernous voids where small items disappear forever, or they're so shallow, you wonder why they bothered. It's a bit like having a pocket that can either hold a full-sized umbrella or a single coin, but nothing in between.

In summary, navigating the interior of the Cybertruck is like wandering through an M.C. Escher painting. It's a bewildering array of design choices that challenge your perceptions of what a vehicle interior should be. It's unconventional, confusing, and occasionally frustrating, but it's never, ever boring. The Cybertruck doesn't just transport you physically; it takes you on a wild ride through the outer limits of automotive design, where functionality meets fantasy, and comfort takes a back seat to style. It's an experience, an adventure, and a puzzle all wrapped up in a stainless steel enigma. And like any good puzzle, once you solve it, you can't help but feel a tiny bit smug.

## Instruments and Guesswork

Delving into the instruments of the Cybertruck is like trying to understand the control panel of an alien spacecraft. You're not sure whether to be amazed or confused, but you're certainly not going to be bored. The instruments in this rolling steel enigma are a combination of high-tech wizardry and what I can only assume is a practical joke by Tesla's designers.

First, there's the speedometer, or what I like to call the 'guess-o-meter'. It's digital, because of course it is, and it displays your speed with the kind of precision that's only useful if you've never heard of speed limits. The numbers change with such rapidity, it's like they're trying to escape from the screen. You're never entirely sure if you're doing 30 mph or warp speed.

Then there's the battery range indicator. This little readout is more optimistic than a lottery player. The range it displays is a fanciful estimate, like the cooking time on a frozen pizza. It's not what's actually going to happen, it's what Tesla would like to happen under ideal conditions – conditions that I suspect include going downhill, with a tailwind, on a road made of silk.

Let's move on to the touchscreen, the pièce de résistance of the Cybertruck's instruments. This screen is to a regular car's infotainment system what a smartphone is to two tin cans connected by a string. It controls everything – and I mean everything. Want to adjust the mirrors? Use the screen. Change the radio station? Screen. Alter the suspension? Screen. It's like Tesla decided that physical buttons were passé and replaced them with a digital overlord.

The navigation system deserves a special mention. It's like having a mildly confused but well-meaning friend who's read a map once. The directions are generally in the ballpark of where you want to go, but you can't shake the feeling that the system is just guessing and hoping for the best. It's the digital equivalent of a shrug.

And then there's Autopilot, Tesla's semi-autonomous driving system. Calling it 'Autopilot' is a bit like calling a dog's lead a 'self-walking kit'. Yes, it can steer, accelerate, and brake on its own, but it requires the same level of trust and vigilance as leaving a toddler with a paintbrush in a white room.

The climate control system is another marvel of guesswork. Adjusting the temperature is like playing a slot machine; you might get what you want, but more likely, you'll just keep trying until you give up. The air vents are hidden, which adds to the minimalist aesthetic, but also means that finding where the air is coming from is like playing a game of thermal hide and seek.

The audio system, meanwhile, is fantastic. It's like having a symphony orchestra in your car, if the orchestra was occasionally interrupted by a DJ who only plays techno. Adjusting the sound settings requires a degree in acoustics and the patience of a saint.

Now, about the vehicle settings. The Cybertruck offers a range of driving modes, from 'Eco' to 'Ludicrous'. Choosing a driving mode is like choosing a mood for your truck. 'Eco' is the Cybertruck trying to be frugal and sensible, like a sumo wrestler deciding to take up ballet. 'Ludicrous', on the other hand, is pure, unadulterated madness, like strapping a rocket to a roller skate.

The rearview camera is a blessing, given that looking out the back of the Cybertruck is like trying to peer through a keyhole. But relying on it is like relying on a periscope in a submarine; it shows you what's happening, but you're never quite sure if it's the whole picture.

Finally, we have the various warning lights and messages that pop up on the dashboard. These range from the mundane – 'Door Open' – to the cryptic – 'Reduced Power Mode'. Deciphering these messages requires a combination of intuition, guesswork, and a willingness to consult This manual, which is, of course, also on the touchscreen.

In summary, the instruments of the Cybertruck are a blend of high-tech sophistication and bewildering guesswork. They provide you with all the information you need, and a lot that you don't, in a format that's as intuitive as a Rubik's Cube. It's an exercise in technological overkill, like using a chainsaw to cut a cake – impressive, but probably more than was strictly necessary.

## Confusing Control Panel Arrangement

Welcome to the befuddling world of the Cybertruck's control panel, a place where logic takes a backseat and confusion reigns supreme. Tesla, in their infinite wisdom, have decided that the traditional control panel layout is outdated, much like my wardrobe according to my daughter. Instead, they've created a control panel that resembles the cockpit of a spacecraft more than something you'd find in a vehicle meant for terrestrial travel.

First, there's the steering wheel, or as I like to call it, the steering rectangle. This geometric oddity looks like it was inspired by a video game console rather than anything you'd expect to find in an automobile. It's as if Elon Musk asked a five-year-old to draw a steering wheel and then thought, "Yes, that'll do nicely." Using this rectangle to actually steer the vehicle is an experience akin to trying to play a piano while wearing boxing gloves - it can be done, but it's not pretty.

Then, you have the massive central touchscreen, the crowning glory of the control panel. This thing is so big, you half expect it to have its own gravitational pull. It controls nearly every aspect of the vehicle, from the climate control to the suspension settings. Navigating through its menus is like trying to solve a Rubik's Cube blindfolded. You know there's a logical way to do it, but that doesn't make it any less infuriating.

The touchscreen's user interface appears to have been designed by someone who thought that minimalism meant removing all helpful cues. Finding the option you need is like playing a game of digital Whack-a-Mole. Just when you think you've found the right menu, it disappears, replaced by something entirely unrelated.

Below the touchscreen, you'll find a sparse array of physical buttons. These buttons are so flush with the surface and unobtrusively designed, you might not even realize they're there. It's like Tesla took the concept of 'sleek design' and ran with it, straight over a cliff. When you do finally locate these elusive buttons, pressing them provides all the tactile satisfaction of poking a ghost.

The indicators and wiper controls have been integrated into the steering rectangle, because why have a simple stalk when you can add more complexity? Using them involves a series of taps and swipes, making you long for the days when a simple flick of the wrist was all that was needed. It's as intuitive as trying to thread a needle while riding a unicycle.

The gear selector, a critical component in any vehicle, is a small, nondescript lever attached to the steering column. It's so unremarkable, you might mistake it for an afterthought. Selecting your desired gear involves a delicate touch, as the difference between 'Drive' and 'Reverse' is a matter of millimeters. It's the automotive equivalent of performing keyhole surgery with a pair of garden shears.

Then there's the volume control, which for some unfathomable reason, is also integrated into the steering rectangle. Adjusting the volume involves a dance of taps and swipes that makes you feel like you're trying to crack a safe rather than simply turning up the radio. It's a task that requires the precision of a bomb disposal expert and the patience of a saint.

The climate control vents are hidden, adding to the minimalist aesthetic, but also ensuring that adjusting the airflow is a game of guesswork. You fiddle with the controls on the touchscreen, hoping for a cool breeze on a hot day, but more often than not, you're just as likely to inadvertently create a miniature tornado in the cabin.

And finally, we have the various warning lights and symbols that flash up on the dashboard. These range from the obvious – like the seatbelt warning – to symbols that look like they were borrowed from ancient hieroglyphics. Deciphering their meaning often requires a degree in cryptology and a healthy dose of guesswork.

In conclusion, the control panel of the Cybertruck is a masterpiece of confusion. It's as if Tesla set out to reinvent the wheel and ended up with a hexagon. Navigating its labyrinthine complexity is an adventure in itself, one that requires intuition, patience, and perhaps a touch of madness. It's not just a control panel; it's a puzzle, a riddle wrapped in an enigma, served with a side of frustration. But once you've mastered it, you can't help but feel a bit like a rocket scientist – albeit one who's still trying to figure out how to turn on the windshield wipers.

# The Maze of Buttons

Step into the Cybertruck and you're immediately confronted with a dashboard that looks like it was lifted straight out of a spaceship from a low-budget science fiction movie. The designers at Tesla, in a fit of what I can only assume was madness induced by too much caffeine, decided that a conventional dashboard was too mundane. Instead, they've given us a control panel that would have even the crew of the Starship Enterprise scratching their heads.

First, let's discuss the steering wheel - sorry, I mean the steering yoke. It's not so much a wheel as it is a tribute to every sci-fi movie Elon Musk watched as a child. Steering the Cybertruck with this yoke feels like you're piloting the Millennium Falcon, not making a simple left turn at the traffic lights. It's a bold design choice, but then again, so was New Coke.

The centerpiece of this intergalactic control panel is, of course, the mammoth touchscreen that dominates the dashboard. This screen is so large and bright, you half expect it to start giving you weather updates from Jupiter. Navigating through its menus requires a degree in astrophysics. Every task, from changing the radio station to adjusting your seat, is buried within a labyrinth of submenus that would make Theseus feel uneasy.

Now, onto the buttons, or lack thereof. Tesla, in its quest to turn the Cybertruck into a rolling smartphone, has done away with most physical buttons. What few buttons remain are so subtly integrated into the dashboard that finding them is like playing a game of 'Where's Waldo?' during a blackout. When you do finally stumble upon one, pressing it provides all the tactile satisfaction of poking a marshmallow.

The air conditioning controls are another marvel of over-engineering. They're hidden within the depths of the touchscreen's menus, making adjusting the temperature in the cabin akin to cracking a safe. You'll swipe, poke, and prod at the screen, hoping for a gust of cold air but more likely to accidentally open the sunroof.

Speaking of the sunroof, the control for it is so cleverly disguised, you might accidentally launch a satellite into orbit while trying to let in some fresh air. It's as if the designers thought that making the controls intuitive would be just too... pedestrian.

The gear selector is a small, unassuming lever that looks like it was borrowed from a child's toy. Using it is as much guesswork as it is a deliberate action. You'll move it gently, half expecting the Cybertruck to transform into a robot rather than shift into reverse.

The indicators and wiper controls have been integrated into the steering yoke, because Tesla believes that stalks are so last century. Using them involves a sequence of swipes and taps that makes Morse code seem straightforward. It's like trying to communicate with an alien species — you're never quite sure if you're getting your point across.

Then there are the various warning lights and symbols that flash up on the instrument panel. These icons are a series of cryptic hieroglyphs that require a Rosetta Stone to decipher. When one lights up, you'll spend more time trying to figure out what it means than actually addressing the issue. It's like playing a game of charades with a car.

And let's not forget about the audio system controls, hidden away in the touchscreen's maze of menus. Adjusting the volume or changing a song is a task that requires the precision of a brain surgeon. You'll prod at the screen, hoping for Beethoven, and end up with death metal at full blast.

In summary, the dashboard of the Cybertruck is less a user-friendly interface and more a test of your problem-solving skills. It's like Tesla threw away the rulebook and wrote a new one in a language only they understand. Navigating this maze of buttons, touchscreens, and hidden controls is not just a challenge; it's an adventure. It's the control panel of an alien spacecraft, designed by a team of people who clearly enjoy a good puzzle. But once you've mastered it, you'll feel like you've been initiated into a secret club — the club of Cybertruck drivers who can actually find the demister.

## Knobs and Dials Galore

In the bizarre world of the Cybertruck, the concept of 'one function per knob' is as alien as the notion of a quiet Lamborghini. Tesla, in an apparent effort to outdo every car manufacturer in the history of ever, has decided that simplicity is overrated. Why have a knob that does one thing, when you can have a knob that requires a manual thicker than "War and Peace"?

Let's start with what I affectionately call the 'Mystery Knob'. This knob, located in a place you'd least expect – next to the cupholder, which is, incidentally, capable of holding everything but cups – is a master of disguise. Twist it one way, and it adjusts your seat. Twist it another, and you're suddenly changing the radio station. Press it, and who knows? You might engage the autopilot or order a pizza.

Then there's the 'Multi-Function Dial'. This little gem is located so conveniently out of reach that you need to dislocate your shoulder to use it. Once you've managed to get your fingers on it, the fun really begins. Rotate it, and it scrolls through your phone contacts. Press it, and it toggles between your tire pressure readings and the phases of the moon. It's like playing a slot machine – you never know what you're going to get, but you're sure it won't be what you wanted.

Now, let's talk about the touchscreen, because, in the Cybertruck, everything comes back to the touchscreen. This is not so much a screen as it is a portal to a parallel universe where logic and intuition are turned on their heads. You'll tap to adjust your mirrors, swipe to check your battery range, and pinch to zoom in on the map – which, by the way, sometimes thinks you're in the middle of the Atlantic Ocean.

The volume control deserves a special mention. In a normal vehicle, adjusting the volume is a simple affair. But this is the Cybertruck, where simplicity goes to die. Here, the volume control is a part of the touchscreen.

But it's not just a matter of sliding your finger up or down. Oh no. You must first activate the volume control mode, which requires a sequence of taps, swipes, and possibly a blood sacrifice.

The climate control system is yet another exercise in over-complication. You want to adjust the temperature? Sure, just give us a moment to navigate through six menus, three submenus, and a quick game of Tetris. And the fan speed is adjusted by a dial that also controls the cabin lighting, because why not?

The gear selector, which in any sane vehicle would be a straightforward lever, is in the Cybertruck a multi-functional joystick. It looks like it was borrowed from a 1980s arcade game. Push it forward to go forward, pull it back to reverse, and move it side-to-side to, I don't know, launch missiles?

The steering wheel, which I've already mentioned is not so much a wheel as it is a geometric nightmare, also houses a multitude of functions. You've got buttons to answer your phone, buttons to adjust your cruise control, and buttons whose purpose remains a mystery. It's like playing a game of 'Guess Who?' but with car functions.

And let's not forget about the various warning lights and messages. These aren't just simple indicators like 'Check Engine'. No, these are cryptic messages that would baffle even the Enigma code breakers. 'Energy Consumption Exceeds Norms' – what does that even mean? Am I driving too fast, or have I accidentally activated a time travel function?

In conclusion, the Cybertruck's approach to knobs and dials is a testament to Tesla's philosophy of 'more is more'. Why have simplicity when you can have complexity? Why have one function per knob when you can have twenty? It's a vehicle that doesn't just transport you from A to B; it takes you on a whirlwind tour of every button, dial, and touchscreen option known to mankind. Driving the Cybertruck isn't just a journey; it's an adventure — a confusing, sometimes frustrating, but ultimately unforgettable adventure.

## The Single Button of Uncertain Functions

Dive into the enigmatic abyss that is the Cybertruck's dashboard, and you'll find yourself face to face with the Single Button of Uncertain Functions. This isn't just a button; it's a riddle wrapped in a mystery inside an enigma, and then crammed into a piece of automotive technology. Tesla, in a moment of what can only be described as sheer madness or absolute genius — the line is blurry — decided that what the Cybertruck really needed was a button that leaves you guessing more than a game of Cluedo.

This button, unassuming in its appearance, sits there like an unexploded bomb, full of potential and menace. It's the kind of button that you're afraid to press, but at the same time, can't resist poking just to see what happens. Will it turn on the radio? Adjust your seat? Or perhaps it'll open a portal to another dimension where the Cybertruck is considered a subtle and understated piece of design.

Press the button, and the first thing you'll notice is... well, nothing. It's like those moments in a horror film where the protagonist opens a door, and there's a momentary pause before all hell breaks loose. You'll sit there, holding your breath, waiting for something to happen. And then, when you least expect it, the car's systems spring to life in a way that's both bewildering and slightly terrifying.

Perhaps the headlights will start flashing in a pattern that suggests you're trying to communicate with extraterrestrials. Or the air conditioning will suddenly blast at full power, turning your vehicle into a mobile Arctic expedition. The possibilities are endless and equally unpredictable.

One day, the button might decide to alter your navigation system, setting a new course to a destination you never knew you wanted to visit. The next day, it might take control of the sound system, blaring out classical music at volumes that would make Beethoven reach for earplugs.

The beauty – or terror, depending on your perspective – of the Single Button of Uncertain Functions is its sheer unpredictability. It's like playing Russian roulette with car features. Each press is a gamble, a roll of the dice, a leap into the unknown. It's automotive anarchy condensed into a tiny, clickable form.

And then, there's This manual. Oh, This manual. Within its pages, you'll find a section dedicated to this button, written in a language that seems to be a cross between technical jargon and the ramblings of a mad poet. The instructions are as clear as mud and twice as thick. They offer no clues, no hints, just a series of cryptic suggestions that leave you more confused than enlightened.

But the true charm of the Single Button of Uncertain Functions lies not in its practicality, but in its ability to surprise and confuse. In a world where everything is labeled, categorized, and explained, this button stands as a bastion of mystery and intrigue. It's a throwback to a time when not everything had to make sense, a time when adventure could be found in the mere press of a button.

So, dear Cybertruck owner, embrace the chaos. Press the button. Revel in the uncertainty. Enjoy the bewildering array of functions that it offers, even if you have no idea what they are. Let the button remind you that sometimes, not knowing is half the fun.

Just be warned: if your Cybertruck suddenly starts to levitate or time travel, don't say I didn't tell you so. Remember, with great power comes great responsibility – or, in the case of the Cybertruck, with a great button comes a great deal of head-scratching.

# Will it launch the car or just the windscreen wipers?

In the labyrinthine confusion that is the Cybertruck's dashboard, there lies a button. Not just any button, mind you, but the Button of Mystery. This button is the automotive equivalent of Schrödinger's cat - you never really know what it's going to do until you press it. It's a gamble, a roll of the dice, a game of Russian roulette with your vehicle's functions.

This button, unlabelled, enigmatic, sits there like a challenge to your curiosity and common sense. It could be anything. Maybe it's the launch control, preparing to propel you into the stratosphere at speeds that would make a fighter jet blush. Or perhaps it's just the windscreen wipers, ready to swish back and forth with all the enthusiasm of a metronome on a slow day.

Pressing the Button of Mystery is like opening Pandora's Box. The first time you press it, maybe nothing happens. It's anti-climactic, a damp squib, a fizzled firework. So, you press it again, and this time, your seat starts adjusting itself, folding and unfolding like a possessed deckchair. It's a wild ride, and not the kind you signed up for when you bought what you thought was just a truck.

But the fun doesn't stop there. The next time you press the button, the sunroof might open, letting in the elements. It doesn't matter if it's raining, shining, or if there's a flock of seagulls overhead, ready to take aim at your interior. The Button of Mystery does not discriminate - it simply does.

One day, this button might decide to alter the suspension settings, turning your smooth ride into a journey reminiscent of a voyage on the high seas. You'll be bobbing and weaving, seasick on dry land, cursing the day you decided to play with the Button of Mystery.

Another day, you might press it and find the radio coming to life, blasting out tunes from a genre you didn't even know existed. It's like having your own personal DJ, one with a very eclectic and unpredictable taste in music. You'll be cruising along, listening to classical, when suddenly it's 1980s punk rock, and you're in a mosh pit on wheels.

And let's not forget the possibility that this button could control something as mundane as the interior lighting. One moment, you're in darkness; the next, it's like Blackpool Illuminations in your truck. It's all very dramatic and entirely unnecessary, but then, that's the beauty of the Button of Mystery.

The instruction manual is no help either. It refers to this button in the vaguest terms, with phrases like "multi-functional interface" and "context-sensitive control". It's the literary equivalent of a shrug. You're on your own, left to decipher the whims of this button through trial, error, and a healthy dose of luck.

In the world of the Cybertruck, the Button of Mystery is a symbol. It represents the unknown, the uncharted territories of automotive design. It laughs in the face of convention and dances on the grave of predictability. It's a wild card, a joker in the pack, a reminder that sometimes, it's the journey, not the destination, that matters.

So go ahead, press the button. Embrace the uncertainty. Enjoy the thrill of not knowing whether you're about to launch your car or just clean your windscreen. It's an adventure, a mystery, a journey into the unknown. Just maybe keep a firm grip on your seat - you never know what's going to happen next.

## A game of guessing what the button does each time you press it.

In the surreal, angular world of the Cybertruck, Tesla has reinvented not just the wheel, but also the very concept of button functionality. Here, we find ourselves playing a game I like to call 'A Hundred Guesses', where each press of a button is as unpredictable as a British summer. It's a game, a gamble, a galactic lottery where the stakes are your sanity and the prize is usually bewilderment.

Imagine a button. A simple button, in a truck that looks like it was designed on a particularly edgy day at a geometry convention. You press this button. What happens? The correct answer is: who knows? Tesla certainly doesn't.

It could be anything from adjusting your seat to changing the dimension of time and space. It's Schrödinger's button – every function exists and doesn't exist until you press it.

Let's start with a scenario. You're driving along, and you decide to adjust the air conditioning. You press a button. Suddenly, you're not adjusting the air conditioning; you're initiating the autopilot, and now you're a passenger in your own car, careening towards the unknown. It's an adventure, sure, but perhaps more of an adventure than you bargained for when you just wanted a bit of cool air.

Next, you might think about changing the radio station. You press what you think is the audio button. Oh no, my friend. Now your Cybertruck thinks it's a disco on wheels. The interior lights start a rave, and you're left wondering whether you're driving a car or commanding a party bus.

And heaven forbid you try to open the sunroof on a nice day. You press the button, and instead of a gentle retraction, you find your seat moving. Forward and back, up and down, like a confused carnival ride. You wanted sunshine, but you got an impromptu chiropractic session instead.

Then there's the button you press, hoping to adjust your mirrors. Ah, but Tesla laughs in the face of predictability. Instead of adjusting your mirrors, your truck now decides to show you its party trick by folding in the mirrors completely. Now you're effectively driving blind, with no idea what's beside you – but at least your truck's aerodynamics are slightly improved.

Don't forget the mystery of the temperature controls. One press could mean cooler air; another press might just crank the heat up to 'Sahara Desert' levels. It's a climatic roulette wheel, and you're gambling with your sweat glands.

Let's say you want to engage the windscreen wipers. A simple task in any other vehicle. But this is the Cybertruck. You press the button, and instead of a gentle sweep of the wipers, the truck's exterior lights start a bewildering sequence of flashing. It's less 'clear my windscreen' and more 'signal to Mars'.

And it's not just the pressing; it's the holding, the double pressing, the press and hold – each a variation in the game of 'A Hundred Guesses'. Each combination offers a new surprise, a new unsolicited adventure. It's like playing a slot machine; you pull the lever and hope for the best, but more often than not, you're left scratching your head.

Through all this, the Cybertruck's manual sits in the glove compartment, thick as a Tolkien novel, filled with explanations as clear as mud. It's less of a manual and more of a 'choose your own adventure' book, where each choice leads you further down the rabbit hole of Tesla's whimsy.

In essence, every button in the Cybertruck is a story, a narrative of uncertainty and surprise. It's a game of guessing, a test of your intuition, or perhaps just your patience. Each journey in the Cybertruck is a game of 'A Hundred Guesses', where the only certainty is the thrill of the unknown, the adventure of the unintentional, and the ever-present risk of ending up somewhere you never intended, doing something you never imagined. But then, isn't that the kind of excitement we all secretly crave in the monotonous predictability of daily life? With the Cybertruck, every drive is a roll of the dice in Tesla's grand casino of automotive eccentricity.

## The "What Does This Do?" Holographic Display

When you clamber into the driver's seat of a Cybertruck, it's like entering the bridge of the Starship Enterprise, only less logical. You're immediately confronted with the pièce de résistance of Tesla's foray into the unnecessary: the "What Does This Do?" holographic display. This is not your average display; it's a kaleidoscope of confusion, a symphony of befuddlement.

Picture this: a display that projects in 3D, hovering in the air like a mirage in the desert. It looks impressive, like something straight out of a sci-fi movie where the special effects budget was bigger than the GDP of a small country. But once the initial awe wears off, you're left with one pressing question: what on Earth does it actually do?

The holographic display, in its infinite wisdom, shows you a myriad of information. Speed, navigation, battery level – all floating in front of you like a ghostly apparition. But the clarity ends there. The display changes with the subtlety of a chameleon on a disco dance floor. One moment you're looking at your speed, the next it's the weather forecast on Mars.

Trying to control this display is akin to trying to herd cats. You wave your hands in the air, attempting to swipe through menus and select options. To the casual observer, it looks like you're engaging in a vigorous session of mid-air Tai Chi. The system relies on gesture controls, which sounds futuristic, but in reality, it's about as accurate as trying to thread a needle while wearing boxing gloves on a rollercoaster.

Let's say you want to change the music track. A simple task in most cars, but not in the Cybertruck. Here, you swipe your hand left, and instead of changing the track, you've somehow adjusted the climate control. Now, instead of listening to Led Zeppelin, you're in a Siberian winter, shivering and still listening to a track you're tired of.

And then there's the navigation. You attempt to set a destination, swiping up to engage the map. Suddenly, you're not looking at a map; you're looking at a live feed of a SpaceX launch. Fascinating, yes, but not particularly helpful when you're just trying to find the nearest petrol station – which, incidentally, you don't need because, remember, you're in an electric truck.

The holographic display also shows your vehicle's stats in real-time. It's like having a health check-up while you're driving. You see your battery life draining away, which is about as comforting as watching your phone battery die while you're lost in the wilderness.

Occasionally, the display decides to offer up some helpful driving tips. These are about as useful as a chocolate teapot. It'll suggest things like, "Drive more efficiently," which is Tesla's way of saying, "You're doing it wrong."

But the real party trick of the holographic display is when you engage the autopilot. Suddenly, you're not just looking at a display; you're looking into the soul of the Cybertruck. It shows you a visual representation of the road, complete with other vehicles, pedestrians, and what I can only assume are ghosts or UFOs, because they certainly don't correlate with anything in the real world.

In summary, the "What Does This Do?" holographic display is a marvel of technology, a testament to Tesla's commitment to the unnecessary. It's a confusing, bewildering, and at times, hilariously unhelpful piece of equipment. It turns every drive into a guessing game, a puzzle where the prize is figuring out how to turn off the 3D projection of Elon Musk's latest tweet floating in front of your face. But in its defense, it does look incredibly cool, and let's be honest, that's half the reason anyone buys a Cybertruck. Who cares about practicality when you can live in the future?

# Futuristic Confusion

In the curious world of the Cybertruck, Tesla has decided that the future is now, even if it means baffling every driver in the process. This truck is a smorgasbord of advanced technology paired with a user interface that might as well be in hieroglyphics. It's like being handed the world's most advanced calculator when all you want to do is add two plus two.

The first thing you'll notice inside the Cybertruck is the abundance of screens. Screens everywhere. You'd think you'd walked into Currys. These aren't just any screens; they're sleek, they're shiny, and they're utterly, utterly perplexing. It seems Tesla believes that the future of driving involves poking at various glossy rectangles and hoping for the best.

Let's start with the central touchscreen, the beating heart of the Cybertruck's technological prowess. This screen, vast as the Sahara, controls everything from your sat-nav to your seat warmers. It's like having an iPad on steroids – but without any of the intuitive user interface. Want to change the temperature? Swipe left, tap twice, and recite the alphabet backwards. It's more complicated than trying to order coffee in a hipster café.

Then there's the matter of the digital instrument cluster. In a normal car, this would show you your speed, fuel gauge, and perhaps a little symbol to tell you that, yes, you've left the handbrake on again. But in the Cybertruck, it's like piloting a spaceship. You're presented with a bewildering array of numbers, graphs, and icons that probably make perfect sense to the engineers at SpaceX, but to the average driver, it might as well be in Klingon.

The steering wheel – sorry, steering yoke – looks like it was ripped straight from a Formula 1 car, because who needs practicality? It has more buttons than a remote control, and deciphering their functions is about as easy as reading Morse code in a sandstorm. You'll press one button hoping to activate the cruise control and end up accidentally opening the glove compartment, which, by the way, is also electronically controlled for some reason.

Voice control, Tesla's answer to the multitude of confusing buttons and screens, isn't much better. It's like having a conversation with a particularly obstinate toddler. You ask it to play some music, and it decides to call your ex. You ask for directions to the nearest petrol station – again, forgetting that you're driving an electric truck – and it helpfully opens the sunroof.

And let's talk about the Autopilot, Tesla's pièce de résistance. Engage it, and the truck drives itself, which sounds fantastic until you realize you're putting your life in the hands of what is essentially a very large computer on wheels. It's like being in a video game, except the consequences of a crash are far more significant than starting over from the last checkpoint.

The exterior cameras, designed to give you a 360-degree view around the truck, are a feature straight out of a Bond film. Unfortunately, trying to switch between the various camera views is like playing a game of whack-a-mole. You tap one camera, and another one activates. It's a never-ending game of technological peekaboo.

Then there's the infotainment system, a term I use loosely. It's less 'entertainment' and more 'guess what I'm trying to do now'. The interface is a riddle, wrapped in a mystery, inside an enigma, and then stuffed into a touchscreen. You'll spend more time trying to figure out how to play a song than actually listening to it.

In summary, the Cybertruck is a marvel of futuristic technology that has somehow forgotten that it needs to be driven by humans. Humans who, for the most part, just want to drive from point A to point B without having to engage in a battle of wits with their vehicle. It's like Tesla set out to create the car of the future but forgot to include an instruction manual for the present. The result is a vehicle that's as confusing as it is impressive – a technological wonder that, for all its advancements, still can't make a cup of tea. And at the end of the day, what good is the future if you can't have a decent cuppa?

## Trying to figure out if it's a map, a movie, or an abstract art piece.

In the ambitious, button-laden cockpit of the Cybertruck, Tesla presents you with what can only be described as a digital conundrum. Here you find a screen – no, not just a screen, a digital canvas, where it's hard to tell if you're looking at a map, watching a movie, or gazing upon an abstract art piece created by a computer having an existential crisis.

Let's start with the navigation system. In any normal car, this would consist of a straightforward map with clear directions. But in the Cybertruck, it's as if you've asked Picasso to paint a street guide. The map swirls and morphs, with roads and landmarks appearing as if they're part of a Dali painting. You want to go to the supermarket, but according to this, you're navigating through the fourth dimension, taking a left turn at the edge of reality.

Then there's the infotainment screen, which I suspect was designed by someone who thought that 'user-friendly' meant 'make the user feel like they're deciphering the Enigma Code'. You press what you think is the button to play music, but suddenly you're watching a live feed of a SpaceX rocket launch. Fascinating, yes, but not particularly helpful when you're just trying to find a decent radio station.

The screen's layout changes more often than a chameleon in a discotheque. One moment it's displaying your vehicle's vital statistics, the next it's showing what can only be described as modern art. It's a whirlwind of colors and shapes, moving and shifting in a way that's both mesmerizing and utterly baffling. You're not sure if you should be driving or visiting an art gallery.

The camera feed is another source of bewilderment. It's supposed to give you a clear view of your surroundings, but the array of angles makes you feel like you're in a low-budget spy movie. You switch from the front camera to the rear camera, but for a moment, you're not sure if you're looking at the road or a scene from 'Blade Runner'. The resolution is crystal clear, but what it's showing you is as clear as mud.

And let's not forget about the 'interactive' part of this bewilderment. The touchscreen responds to your inputs in a way that suggests it's just guessing what you want. You swipe left, and it zooms in. You swipe right, and suddenly you're in the settings menu, adjusting the ambient lighting to a shade of neon green that you didn't even know existed.

Interacting with this system feels like you're playing a game where you don't know the rules, and the rules keep changing. It's a constant battle of wits between man and machine, and more often than not, the machine seems to have the upper hand. The screen's response to your touch is either overly enthusiastic, launching apps and opening menus with the slightest brush, or it's like prodding a sleeping bear, unresponsive and grumpy.

Then there's the matter of trying to figure out whether the information displayed is current or a relic from a past journey. The system seems to have a mind of its own, displaying routes you took days ago or suggesting destinations that you're pretty sure don't exist in this dimension.

The voice control system doesn't offer much respite either. It's like talking to a particularly obtuse parrot. You ask for directions to the nearest petrol station, and it cheerfully responds with the weather report for Timbuktu. You ask it to play some classic rock, and it decides this is a great time to catch up on the latest news about the stock market.

In conclusion, the interactive experience in the Cybertruck is akin to entering a digital labyrinth with no map and a blindfold. It's a mix of high-tech wizardry and baffling design choices, where every interaction is a roll of the dice. You're never quite sure if you're controlling the truck, or if it's taking you on a wild ride through the digital cosmos. But, amidst all this bewilderment, one thing's for sure: it's never dull. Driving a Cybertruck is an adventure, a journey into the unknown realms of interactive technology, where the line between utility and artistry is blurred into oblivion.

# 3. Starting and Stopping

STARTING AND STOPPING

In the realm of Tesla's Cybertruck, the acts of starting and stopping are as straightforward as explaining cricket to an American. When you enter this polygonal behemoth, don't expect a simple turn of the key. No, that would be too easy, like making a sandwich. Here, you're solving a Rubik's Cube blindfolded.

To start the Cybertruck, first, you must find the 'start' button. This is Tesla's little game of hide and seek. The button is camouflaged better than a chameleon in a bag of Skittles. It's subtly integrated into the dashboard, which is as minimalist as a monk's living quarters. After pressing several buttons that could either adjust your seat or launch a satellite, you'll accidentally press the correct one and the truck awakens with all the drama of a library.

Once the truck is on, it's time to select your gear. But in the Cybertruck, the gear selector is like a riddle wrapped in a mystery inside an enigma. It's a stalk on the steering column that seems to have a mood disorder. Nudge it gently, and you might end up in 'Drive'. Push it with a bit more enthusiasm, and suddenly you're reversing, rapidly. It's less about precision and more about good luck and a fair wind.

Now, let's talk about stopping. In a normal vehicle, you'd gently apply the brakes. But in the Cybertruck, the brakes are as sensitive as a poet on Valentine's Day. Press them too gently, and you'll continue to sail forward with all the grace of an ocean liner. Press them too hard, and you'll jolt to a stop, throwing everything and everyone forward in a beautiful demonstration of Newton's first law.

And here's a fun fact: the Cybertruck's brakes regenerate power back to the battery. It's like the truck is saying, "Thank you for stopping; here's some electricity." It's a nice idea, but it also means the brake pedal feels different each time you press it. It's not a pedal; it's a mystery wrapped in rubber.

When parking this electric leviathan, you're engaging in a task comparable to docking the International Space Station. You need the spatial awareness of a bat and the foresight of a chess grandmaster. The Cybertruck doesn't just park; it claims territory.

The parking brake is another adventure. It's not a lever you pull with a satisfying ratcheting sound. No, it's a button, another button in a sea of buttons. You press it and hope it's the parking brake and not something that adjusts your seat to a chiropractic nightmare position.

And when it's finally time to turn off the Cybertruck, you'll go through the reverse of your starting procedure. This involves pressing the start button again, which now decides it's the stop button. But here's the twist: sometimes the Cybertruck doesn't shut down immediately. It lingers, like a guest at a party who doesn't realize it's time to leave. The screens stay on, the lights stay bright, and you sit there, wondering if it's waiting for a goodnight kiss.

In summary, starting and stopping the Cybertruck is an experience akin to a dance, a tango with technology where you're never quite sure who's leading. It's a ballet of buttons, a waltz of wheels, a pas de deux of power and perplexity. But amidst this dance, you find the essence of the Cybertruck – a vehicle that refuses to do anything ordinarily, a testament to Tesla's philosophy that even the simplest actions deserve a sprinkle of complexity and a dash of drama. So, buckle up, press the right button (hopefully), and enjoy the ride – it's going to be unique.

## The Guessing Game of Engine Ignition

In the mystifying world of the Cybertruck, starting the engine – or rather, engaging the electric motors – is akin to participating in a game show where you're always the contestant and never the host. The simple act of igniting a traditional engine is replaced by an enigma, wrapped in a conundrum, shrouded in mystery.

Firstly, let's establish one thing: there is no key. The concept of a physical key is as archaic in the Cybertruck as a hand-cranked engine. Instead, you get a fob – a sleek, mysterious piece of technology that looks like it could double as a futuristic bottle opener. This fob doesn't have buttons. Oh no, that would be too pedestrian. It communicates with the truck telepathically or via dark sorcery – the jury's still out on that one.

Upon entering the Cybertruck, you're not greeted by the familiar rumble of an engine coming to life. Instead, the interior lights up like a Christmas tree in Times Square. It's Tesla's way of saying, "Yes, I'm awake, what do you want?" The dashboard, a minimalist expanse of screens, comes to life, displaying an array of information that's as comprehensible as quantum physics.

Now, to actually 'start' the Cybertruck, you must engage in the guessing game. There's a button. Of course, there's a button. But it's not labeled. Why would it be? That would make things too easy. This button is flush with the dashboard, camouflaged better than a chameleon in a jungle. You press it, half-expecting to eject yourself through the roof. But no, instead, the display merely flickers, as if mocking your attempt.

After several attempts of pressing what you think might be the start button – accidentally turning on the windscreen wipers, the hazard lights, and what might be the truck's self-destruct mechanism – the Cybertruck finally decides to acknowledge your efforts. The electric motors whirr silently to life, as undramatic as a librarian reshelving a book.

You see, in the Cybertruck, there is no roar, no satisfying growl of an engine. It's more of a whisper, a subtle hint that something might be happening. It's about as exciting as watching paint dry, but with less smell.

And don't think the guessing game ends once you've started the truck. Oh no, it's just beginning. Now you must figure out how to get it into gear. You'll find a stalk on the steering column. It looks innocuous enough, but it holds power over your entire driving experience. Nudge it gently, and you might find yourself in drive. Nudge it too enthusiastically, and you'll be reversing faster than you can say, "Oops."

This stalk doesn't believe in clear instructions. It operates on a level of intuition that would baffle a psychic. You think you're telling it to go forward, but the Cybertruck has other ideas. It's less of a gear selector and more of a suggestion box.

In conclusion, starting the Cybertruck is not just an action; it's an experience, a journey, a guessing game of technological wits. Each time you enter the truck, you're not just a driver; you're a contestant in Tesla's game of 'Guess How to Start Me.' It's an adventure in modern motoring, a foray into the future of transportation, where the simple turn of a key is replaced by a series of gambles with a computer on wheels. Welcome to the future; it's confusing, but at least it's quiet.

## Is today the day it decides to start, or will it just make whirring noises?

In the ever-bewildering universe of the Cybertruck, starting the vehicle is less a daily routine and more akin to spinning the wheel on a game of Russian roulette. Each attempt to awaken this electric beast is fraught with the suspense of a Hitchcock thriller and the unpredictability of a British summer.

Upon entering the Cybertruck, which bears more resemblance to an escape pod from a sci-fi film than any car, you're first tasked with deciphering the means to start it. Traditional keys are for people who still believe in hand-cranking their engines. Here, you have a key fob that's so sleek and minimalistic, it's practically useless. It doesn't have buttons – that would be too conventional. Instead, it communicates with the truck in a series of cryptic winks and nudges.

Approaching the vehicle, you wave the fob in the general direction of the truck, like a wizard casting a spell. The doors unlock, which is a promising start. Now, for the main event: starting the truck. You're greeted by a dashboard that looks like it was designed by someone who thinks buttons are a blight on humanity. The start button is hidden, part of Tesla's ongoing commitment to turning simple tasks into a treasure hunt.

You press the button, or at least where you think the button is. The truck responds with a series of lights and sounds, like R2-D2 having a nervous breakdown. But does the engine start? Not yet. It's contemplating. This is the Cybertruck's way of building suspense, a tribute to the days of yore when engines had a crank and starting a car was a workout.

On a good day, the press of the button is met with the gentle hum of the electric motor, as silent and smooth as a cat stalking its prey. But on other days, it's more like a game of ignition roulette. You press the button, and the truck responds with an orchestra of whirring noises, beeps, and flashing lights. It's as if the truck is trying to communicate in Morse code.

Sometimes, the Cybertruck decides it's not quite ready to face the world. The dash lights up, the screens come to life, but the motor remains in a deep slumber. It's like trying to wake a teenager on a school day – you know it has to happen, but it takes an age and a considerable amount of prodding.

And let's not forget the times when the Cybertruck decides to throw a curveball. You press the start button, and instead of the gentle hum of the motor, you're greeted with the infotainment system launching into an impromptu concert. The speakers blare at full volume, giving you a heart attack and a half, but as for moving – the truck remains steadfastly stationary.

Then there's the ever-present threat that the Cybertruck, in its infinite wisdom, will choose to update its software just when you need to start it. The message "Installing Updates" appears on the screen, and you're trapped in a purgatory of progress bars and loading icons. It's like waiting for a kettle to boil, but less predictable and far more frustrating.

On the occasions when the stars align and the Cybertruck deigns to start, there's a palpable sense of achievement. It's not just starting a car; it's winning a battle of wits against a vehicle that's arguably smarter than you. You drive away not just with a sense of destination, but with a feeling of triumph.

In conclusion, starting the Cybertruck is an exercise in patience, luck, and keeping your sense of humor intact. It's a daily game of ignition roulette, where each start is a gamble, and the odds are never quite in your favor. But then again, who needs predictability when you have a truck that looks like it could survive an apocalypse? So, take a deep breath, press that button, and hope for the best. After all, it's not just a drive; it's an adventure.

# The Key to Nowhere

In the bewildering and somewhat nonsensical world of the Cybertruck, the key fob stands as a monument to Tesla's unending quest to complicate the simple things in life. This isn't just a key fob; it's a riddle, wrapped in a mystery, inside an enigma, and then dipped in a healthy dose of confusion for good measure.

First, let's address the design. Tesla, in what can only be described as a moment of madness, decided that the key fob should not resemble anything remotely key-like. Instead, they've given you a sleek, smooth piece of metal that looks more like a minimalist art piece than something that's supposed to unlock a car. It's the kind of object you'd find in a modern art gallery, with a plaque underneath that reads "The Essence of Emptiness."

Now, to the functionality, or lack thereof. Traditional key fobs have buttons – lock, unlock, maybe a button to open the boot, and if you're feeling particularly adventurous, a panic button. But Tesla scoffs at such pedestrian ideas. Why have buttons when you can have a fob that responds to telepathic commands or interpretive dance?

You approach the Cybertruck, key fob in hand, ready to unlock it. But how? There are no buttons. You wave the fob around like a magic wand, hoping for the best. Sometimes, the truck unlocks. Other times, it sits there, unmoved by your attempts at vehicular communication, as stoic as the Queen's Guard.

Then there's the issue of actually starting the truck with this fob. In any normal vehicle, you insert the key, turn it, and the engine roars to life. In the Cybertruck, you place the fob in a specific, unmarked spot on the dashboard and hope that it decides to acknowledge your presence. It's less starting a car and more participating in a séance.

Sometimes, the fob decides it doesn't want to play ball. You place it in the spot, and nothing happens. No lights, no gentle hum of the electric motor – nothing. It's as if the fob and the truck are having a silent argument, and you're caught in the middle, an awkward bystander to this technological tiff.

But let's not forget the times when the key fob performs its party trick – opening the windows. Yes, for some reason, Tesla thought that what drivers really need is a way to accidentally open their windows. You put the fob in your pocket, and as you walk away, you press against it just right, and voilà, your windows are down, and the interior is now enjoying the great British weather.

And charging the fob? Oh, it needs charging. Because why have a simple, replaceable battery when you can have yet another thing to charge? It's like having a pet – you need to feed it, nurture it, and make sure it's charged, or else it decides to take a nap at the most inopportune moment.

In conclusion, the key fob for the Cybertruck is not just a key fob; it's an exercise in patience, a test of your problem-solving skills, and a practical joke wrapped in brushed aluminum. It's the key to nowhere, the enigma of access to the Cybertruck, a testament to the fact that in the world of Tesla, even the simplest things are needlessly complicated. But then again, who wants a boring old key when you can have a piece of the future in your pocket? Just remember to charge it, don't lose it, and maybe carry a traditional key as a backup, just in case.

## Asphalt Ambiguity vs. Off-Road Roulette

In the Cybertruck, selecting a driving mode is less about making a choice and more about embarking on a journey of uncertainty. It's akin to choosing your meal at a restaurant with no menu and a chef who's a culinary enigma. You have two primary choices in this electric monstrosity: Asphalt Ambiguity for the tarmac and Off-Road Roulette when you feel like getting lost in nature - potentially quite literally.

Asphalt Ambiguity, the so-called standard driving mode, is about as standard as finding a vegetarian shark. It's supposed to be your go-to for everyday driving, but it's more like playing a game of guess-the-gear. You expect smooth acceleration and civilized road manners. Instead, the Cybertruck surges forward with the eagerness of a greyhound that's just spotted a rabbit. It's less of a gentle cruise down the motorway and more of a white-knuckle ride through Tesla's interpretation of what roads should feel like.

In this mode, the suspension decides it's a ballerina and prances over every bump and dip. It's so responsive that you feel every pebble, every imperfection on the road. It's as if the truck is trying to communicate the road's life story through your spine. Comfort is apparently for the weak, and the Cybertruck is here to make men out of its drivers, one pothole at a time.

Then there's Off-Road Roulette, a mode that should come with a disclaimer: "For the brave or foolish." When you select this mode, the Cybertruck raises itself like a cat arching its back, ready to pounce. It's preparing to tackle the wild, or at least a slightly unkempt gravel path.

Engaging Off-Road Roulette is where the real fun begins. The truck's demeanor changes. It becomes a beast, ready to devour the untamed terrain. But here's the catch: it's about as predictable as a game of Russian roulette with the settings. One moment you're smoothly traversing a muddy track, the next you're bouncing around like you're inside a paint mixer. The suspension, so delicate and dainty on the road, now decides it's auditioning for a monster truck rally.

The throttle response in this mode is equally schizophrenic. Sometimes a gentle press propels you forward with the subtlety of a stampeding elephant. Other times, you need to floor it just to get over a small incline. It's a guessing game: Will I gently ascend this hill, or will I launch myself into low Earth orbit?

And let's talk about the traction. The Cybertruck, in Off-Road Roulette mode, has more grip than a politician clinging to power. But this grip comes with a learning curve steeper than the Eiger's north face. The truck grips and grips until suddenly it doesn't, and you find yourself in a ballet of sliding, spinning, and praying to the god of internal combustion engines, despite driving an electric vehicle.

Switching between these modes is done through the infotainment system, a process that's about as straightforward as explaining quantum physics to a toddler. You navigate through menus upon menus, submenus upon submenus. It's a digital labyrinth, and at the center instead of a Minotaur, you find the driving modes, jeering at you.

In conclusion, selecting a driving mode in the Cybertruck is an exercise in faith, a leap into the unknown. Whether you choose Asphalt Ambiguity or Off-Road Roulette, you're not just choosing a driving setting; you're choosing an adventure. It's a roll of the dice, a spin of the wheel in this electric casino on wheels. But no matter the choice, one thing is certain: it will never be dull. It's Tesla's way of reminding you that driving should be an experience, a story, a chapter in the ongoing saga that is owning a Cybertruck. So choose your mode, hold onto your seat, and enjoy the ride. After all, isn't that what driving is all about?

## Deciding whether "Sport" mode actually means "Slightly Less Slow."

In the world of the Cybertruck, where everything is as clear as a foggy day in London, selecting the so-called "Sport" mode is a gamble akin to guessing the number of sweets in a jar at a village fête. The name suggests a jaunt into the realm of increased performance and agility, but in Tesla's universe, "Sport" might as well mean "Slightly Less Slow."

When you engage Sport mode, you expect a transformation, a metamorphosis from a sedate electric mammoth into a nimble gazelle.

What you get, however, is more akin to a mammoth deciding it might jog instead of walk. The change in acceleration is there, yes, but it's subtle – like the difference between English cuisine with salt and without.

You press the accelerator, anticipating the surge of power that the word 'sport' implies. The Cybertruck responds, but it's like waking a sleeping teenager; it takes its time and doesn't seem particularly happy about it. You're thrust back into your seat, but it's more a gentle nudge than the hand of God catapulting you down the road.

Then there's the handling. In Sport mode, you'd expect the Cybertruck to corner like it's on rails, defying the laws of physics with its grace and agility. The reality is that it corners like a barge trying to navigate the canals of Venice. It's less about finesse and more about brute force and ignorance. The suspension stiffens, but it's like stiffening a mattress by putting a book under it – technically more rigid, but not noticeably more comfortable or effective.

Sport mode also tinkers with the steering, giving it a heavier, more responsive feel. It's Tesla's way of saying, "Look, it's sporty!" But in reality, it's like swapping a feather pillow for a slightly less fluffy feather pillow. The difference is there, but it's not going to revolutionize your driving experience.

But let's not forget the noise, or rather, the lack of it. Traditional sport modes are accompanied by the roar of an engine, the symphony of internal combustion. In the Cybertruck, engaging Sport mode is as silent as the rest of the driving experience. It's like putting on a thrilling movie but forgetting to turn on the sound. You see the action, but you don't feel it in your bones.

And then there's the question of efficiency. Engaging Sport mode in a regular car means throwing fuel economy out of the window. In the Cybertruck, it means depleting your battery with the voracity of a child eating sweets. You can almost see the range dropping, mile by mile, like the countdown on a bomb in a spy movie. It's a constant reminder that fun comes at a cost, and in an electric vehicle, that cost is range.

Selecting Sport mode in the Cybertruck is a bit like buying a lottery ticket. You hope for a life-changing experience, but deep down, you know you're just going to end up with the same life, only slightly less money. It's Tesla's way of giving you the illusion of sportiness, a façade of performance in an otherwise sensible vehicle.

In conclusion, Sport mode in the Cybertruck is an exercise in optimistic branding. It promises a thrilling experience, a taste of the racetrack, but delivers something closer to a brisk walk in the park. It's not so much "Sport" as it is "Slightly Less Slow." But then again, when you're driving a vehicle that looks like it was designed for a Mars mission, maybe a gentle jog is all the excitement you really need. After all, who buys a Cybertruck for the sportiness? You buy it because it looks like the Batmobile's environmentally conscious cousin. Sport mode is just a nice little button that lets you pretend, for a moment, that you're driving something fast. It's the automotive equivalent of putting on a superhero cape and running around your garden; you're not actually flying, but it's fun to pretend.

## Taking bets on whether "Off-Road" mode is actually just "Panic" mode.

Embarking on an off-road adventure in the Cybertruck is like playing a game of Russian roulette with the terrain. You see, Tesla's interpretation of 'Off-Road' mode appears to be throwing everything at the problem and seeing what sticks, much like a toddler trying to fit a square peg into a round hole.

When you engage Off-Road mode, you expect to become master of the wilderness, taming the untamed with the flick of a switch. But what actually happens is akin to opening Pandora's box. The suspension, deciding it's audition day for 'Dancing with the Stars', starts to make moves that would put a ballet dancer to shame. You're not so much driving over bumps as you are participating in a vehicular interpretive dance.

The throttle response in this mode is about as predictable as a soap opera plot. Press the pedal, and you might lurch forward with the gentleness of a charging rhino, or you might find the truck responding with all the enthusiasm of a slug on a lazy Sunday afternoon. It's less of a refined tool and more of a blunt instrument – like conducting an orchestra with a sledgehammer.

And let's talk about the traction control system, which, in Off-Road mode, seems to adopt a personality akin to a paranoid meerkat. The slightest slip, the tiniest loss of grip, and it kicks in with the subtlety of a bull in a china shop. It's not so much assisting you as it is arguing with you, a backseat driver that's gained control of the wheels.

The ride height adjusts, giving you enough clearance to tackle what you think might be a small mountain. But in reality, it's about as useful as a chocolate teapot. Sure, you're higher off the ground, but the truck still handles obstacles with the finesse of a drunken elephant. It's like giving a hippo stilts and expecting it to become more agile.

Then there's the steering, which in Off-Road mode becomes as stiff as the upper lip at a British tea party. You wrestle with the wheel, fighting for control as if you're arm-wrestling with the truck. It's supposed to give you a more 'connected' feel to the terrain, but it just ends up feeling like you're trying to steer a stubborn cow.

The screen displays an array of off-road statistics – incline angles, wheel articulation, and so on. It's information overload, like being bombarded with facts by an overenthusiastic tour guide when all you want to do is enjoy the scenery. You're trying to focus on not rolling the truck, but the dashboard insists on telling you that your left rear wheel is three degrees off from where it thinks it should be.

And let's not even get started on the "Off-Road" cameras. They're supposed to give you a better view of the obstacles around you. In reality, it's like trying to navigate using a kaleidoscope. The cameras show rocks and ruts in such high definition that you start to feel every bump and dip in your very soul.

Choosing to engage Off-Road mode in the Cybertruck is a gamble, a roll of the dice where the odds are about as clear as mud. It's a mode that seems to have been designed by someone who believes off-roading is just driving on slightly uneven pavements. It's less about conquering the wild and more about holding on for dear life while the truck decides it's a monster truck.

In conclusion, Off-Road mode in the Cybertruck is a mixture of optimism and overkill. It's like putting on a suit of armor to go for a stroll in the park – technically it's more protection, but you're going to look and feel a bit silly. It's a mode that's less about providing a tailored off-road experience and more about giving you a story to tell, assuming you make it back in one piece. It's not so much 'Off-Road' mode as it is 'Panic' mode – panic for you, panic for your passengers, and panic for any wildlife unfortunate enough to be in the vicinity. But, at the end of the day, it's all part of the adventure that is owning a Cybertruck. Just remember, when you press that button, you're not just going off the beaten path, you're paving a new one – whether you like it or not.

## Autopilot: Trusting Your Life to Software

In the futuristic world of the Cybertruck, Tesla has generously provided an Autopilot feature, which essentially allows you to entrust your life to a piece of software that, for all you know, could have been programmed by a caffeine-addled intern on a deadline. The concept of Autopilot in a vehicle that looks like it was designed for a lunar expedition is akin to asking a toaster to make you a cup of tea; it's ambitious, but you're not quite sure it's going to work out as intended.

Engaging the Autopilot is like handing over the keys of your house to a stranger and hoping they don't redecorate. You press a button, and the truck takes over, steering, accelerating, and braking with the confidence of a teenager who's just passed their driving test. It's all very impressive until you remember that you're in a several-ton electric behemoth that could flatten a garden shed if the software so much as glitches.

The Autopilot's understanding of road markings is reminiscent of a child's interpretation of a coloring book – it knows where the lines are, but it doesn't always stay within them. You'll find yourself gripping the steering wheel, not because you need to steer, but because it's the only thing keeping you from having a full-blown panic attack as the truck casually drifts towards the lane marker.

Then there's the obstacle detection, which in theory, is brilliant. The truck can supposedly identify everything from a rogue shopping cart to a small dog wandering down the road. In practice, however, it's more like playing a game of 'I Spy' with someone who's forgotten their glasses. Sometimes it'll spot a plastic bag floating across the highway and slam on the brakes as if it's a boulder. Other times, it'll remain oblivious to a pothole the size of a swimming pool until you're airborne.

Using Autopilot in traffic is another gamble. It handles stop-and-go traffic with all the finesse of a ballroom dancer with two left feet. One moment you're crawling along smoothly, the next you're lurching forward like a drunk uncle at a wedding. It's less of a relaxing drive and more of a lesson in unpredictability.

And of course, there's the legal disclaimer. Tesla, in a stroke of genius, reminds you that you need to keep your hands on the wheel and your attention on the road, even when Autopilot is engaged. This effectively turns drivers into glorified supervisors of a student driver, except the student is a computer with an overconfidence complex.

The real fun begins when Autopilot encounters a situation it wasn't programmed to handle. A closed lane, an unexpected detour, or heaven forbid, a roundabout. In these moments, the Autopilot reacts with all the calmness of a cat in a bathtub. It either hands control back to you with a digital shrug, or it makes a decision that could best be described as 'creative', leaving you to handle the aftermath.

In conclusion, using Autopilot in the Cybertruck is like playing a video game with the difficulty set to 'I'm feeling lucky'. It's a technological marvel, a glimpse into a future where cars drive themselves, and humans are just along for the ride. But in its current state, it's like trusting a very smart, very enthusiastic, but ultimately inexperienced teenager with your car. It can be a godsend, a nightmare, or a rollercoaster ride – sometimes all within the same journey. So sit back, relax, and enjoy the future of driving, just don't take your hands off the wheel, your eyes off the road, or your heart rate for granted.

## A Leap of Faith

In the grand, futuristic tapestry that is the Cybertruck, Tesla invites you to place your faith in a system that has all the reliability of a chocolate teapot in a heatwave. This system, a marvel of modern technology, is about as dependable as my decision-making skills after three pints.

Let's start with the navigation system. In theory, it should guide you effortlessly to your destination, as smoothly as a butler carrying a tray of champagne glasses. In practice, it's more like being directed by a slightly inebriated pirate with a penchant for scenic routes. The system's idea of the 'best route' often includes detours through neighborhoods you didn't know existed, leading you on a magical mystery tour that's about as magical as finding out your lottery ticket is a dud.

Then there's the vaunted voice command feature, a system that promises to obey your every verbal whim. In reality, it has the comprehension skills of a toddler. You ask it to play some rock music, and it responds by adjusting the climate control. It's like having a conversation with someone who's only ever read about human interaction in a book.

The lane-keeping assist is another exercise in blind faith. It's supposed to keep you neatly between the lines, as precise as a Swiss watch. But it behaves more like a drunk uncle at a wedding, veering from side to side with an alarming lack of self-awareness. You activate it hoping for a helping hand, but end up gripping the steering wheel like it's the only thing keeping you tethered to this world.

And let's not forget the adaptive cruise control. This feature is designed to maintain a safe distance from the car in front, slowing down and speeding up as necessary. However, its idea of a 'safe distance' is as variable as the British weather. One moment it's tailgating like a teenager in a hot hatch, the next it's leaving enough space for an articulated lorry to slot in, inviting other drivers to cut in front of you with the enthusiasm of shoppers at a Black Friday sale.

The collision avoidance system deserves a special mention. It's intended to be your guardian angel, stepping in to prevent accidents. In practice, it's more like a paranoid bodyguard, overreacting to every perceived threat. A leaf blows across the road, and the system slams on the brakes as if it's a boulder rolling down a hill. It's less about avoiding collisions and more about testing the strength of your seatbelt and the fortitude of your nerves.

Parking assist is another gem in the crown of Tesla's technological achievements. It should make parking a breeze, a simple push of a button and the car elegantly glides into the space. Instead, it's like playing a game of chance. Will it neatly park the car, or will it inch forward, hesitating like a nervous learner driver before ultimately giving up and leaving you half in, half out of the space, much to the amusement of onlookers.

In summary, trusting the Cybertruck's various systems is a leap of faith, a dive into the unknown, a test of your courage, and your patience. It's like relying on a weather forecast in the UK; you know it's probably wrong, but you cling to hope. Each journey in the Cybertruck becomes an adventure, a roll of the dice in the casino of automotive technology. You might reach your destination unscathed, or you might end up taking an unexpected tour of the countryside, serenaded by the air conditioning system, with your adrenaline levels slightly higher than recommended. But then again, who buys a Cybertruck for a mundane driving experience? In the world of Tesla, it's not the destination that matters; it's the unpredictably thrilling journey. So buckle up, take a deep breath, and enjoy the ride – just maybe keep a map and a sense of humor handy.

## The Art of Letting Go

In the curious and often bewildering experience of driving a Tesla Cybertruck, one must master the art of letting go, particularly when the truck, with its Autopilot feature engaged, seems to have a magnetic attraction to every ditch, pothole, and hedge it encounters. It's like teaching a toddler to walk; they sort of know what they're doing, but you can't help but feel they might faceplant at any moment.

When you first activate the Autopilot, a sense of unease washes over you. It's like handing over the controls of your life to a Labrador with a driving license. There's an excitement mixed with a profound sense of dread. The truck veers towards the edge of the lane, eyeing up the ditch like it's spotted an old friend. You grip the steering wheel with white-knuckled terror, ready to take over at the first sign of an impromptu off-roading adventure.

The Cybertruck, in its infinite wisdom, decides that the middle of the lane is merely a suggestion rather than a rule. It's like watching a drunken bumblebee's flight path. There's a constant, nagging fear that you're only a small software glitch away from becoming part of the scenery. It's an exercise in trust, like letting a blind man guide you across a busy road.

And then there's the moment when a bend in the road appears. The Cybertruck approaches it with the overzealous enthusiasm of a puppy chasing its tail. You find yourself muttering silent prayers to deities you don't believe in, hoping that the truck knows what it's doing. The bend is navigated with all the precision of a sledgehammer cracking a nut. It's not so much a smooth arc as it is a series of minor corrections that sees you ping-ponging down the road.

Relaxing in these conditions is like trying to meditate in a room full of mosquitoes. You know you should be calm, that the myriad of sensors and cameras are keeping an electronic eye on things. But the primal part of your brain, the part that's aware you're hurtling through space in a metal box at 70mph, is screaming that this is not natural.

Occasionally, the Autopilot's sensors will detect something. What that something is, nobody knows. It could be a shadow, a low-flying bird, or perhaps a ghost. Whatever it is, the Cybertruck reacts by either slamming on the brakes or swerving slightly, just to keep you on your toes. It's like having a nervous learner at the wheel, only this learner weighs several tons and is powered by enough electricity to light up a small village.

You tell yourself that this is the future, that this is what progress looks like. But as the truck decides to get up close and personal with a hedge on a whim, you can't help but feel that progress might be overrated.

Engaging in conversation or trying to listen to music while the Cybertruck dances with ditches is like trying to maintain a train of thought in a room full of screaming toddlers. You start a sentence, only to abort it with a strangled yelp as the truck makes another lunge for the undergrowth.

In conclusion, learning to relax in the Cybertruck when it seems hell-bent on exploring every ditch in the countryside is an art form. It's a practice in patience, a test of your stress levels, and a reevaluation of your life choices. But there's a silver lining – it's never boring. Every journey is an adventure, a story to tell, one that usually ends with "And then I thought we were going to die." So sit back, try to relax, and remember – the future is here, and it's terrifyingly exciting. Just maybe keep one hand on the steering wheel, just in case.

# 4. Discomfort Systems

In the grand and often puzzling world of the Tesla Cybertruck, the notion of 'comfort' is redefined in ways that would make even the Marquis de Sade raise an eyebrow. The Cybertruck, with its array of 'Discomfort Systems', seems to have been designed with the philosophy that to truly appreciate comfort, one must first experience its exact opposite.

Let us begin with the seats, which Tesla claims are ergonomically designed. This must be a new definition of 'ergonomic' that I wasn't previously aware of. Sitting in these seats is akin to perching on a sculpture in an art gallery – it looks fantastic, but five minutes in and you're experiencing a level of discomfort that can only be likened to medieval torture devices. The lumbar support, in particular, seems to have been designed by someone who confused the spine with a railroad track.

Moving on to the suspension system, the Cybertruck takes the idea of a 'firm ride' to new, spine-jarring levels. Every bump, pothole, and tiny pebble on the road is transmitted directly to your posterior with the efficiency of a telegraph operator. It's like riding in a horse-drawn carriage, minus the horse and on a cobblestone street. You don't so much drive over roads as you interpret Morse code messages from them.

The climate control system, or what I've affectionately dubbed 'Guess the Temperature', deserves a special mention. The Cybertruck, in a bold move, does away with the conventional wisdom of dials and switches. Instead, you're presented with a touchscreen that requires a combination of intuition, luck, and perhaps a degree in meteorology to operate. Adjusting the temperature is like playing a slot machine – you might pull the lever and get a blast of Arctic air, or you might get a gust of Saharan heat.

Then there's the in-cabin noise, or lack thereof. Electric vehicles are known for their quiet operation, but the Cybertruck takes this to an extreme. It's so silent inside the cabin that you can hear your own heartbeat – and the increasingly frantic beat of your heart as you attempt to navigate the truck's various systems. This level of silence is unnerving, creating a sensory deprivation chamber on wheels where even the softest whisper sounds like a shout.

The infotainment system, a centerpiece of the Cybertruck's interior, is another contributor to its discomfort systems. The screen is vast, which sounds good on paper, but in practice, it means taking your eyes off the road to fiddle with settings. It's like trying to watch TV and drive at the same time, except the TV is also your control panel, and you can't find the remote.

Let's not forget the vehicle's handling. The Cybertruck, with its bulky frame and hefty weight, handles corners with all the grace of an elephant on a unicycle. You approach each turn with a sense of trepidation, wondering if this is the moment physics decides to take a holiday and leave you in a ditch.

In conclusion, the Cybertruck's array of Discomfort Systems is a masterclass in how to take the concept of luxury and turn it on its head. It's a vehicle that challenges your notions of comfort, pushing you into realms of experience you never thought possible, primarily discomfort and mild bewilderment. It's not just a truck; it's an endurance test, a trial by fire, a journey into the unknown realms of discomfort. But then again, who buys a Cybertruck for a cushy ride? You buy it because it looks like it could survive an apocalypse, not because it feels like riding on a cloud. In the world of the Cybertruck, comfort is overrated. Discomfort, on the other hand, is an adventure.

## Unpredictable Climate Control Shenanigans

In the fantastical world of the Cybertruck, the climate control system is less a feature and more a whimsical entity, capable of mood swings that would make a soap opera character seem stable. You don't so much set the temperature as you engage in a game of thermal roulette, where the only certainty is your eventual confusion and discomfort.

The moment you attempt to adjust the climate control in the Cybertruck, you realize that Tesla must have misunderstood the term 'user-friendly'. The controls are buried in the depths of the infotainment system, requiring a series of swipes and taps reminiscent of trying to crack the Enigma code. The process of changing the temperature is akin to trying to solve a Rubik's Cube while wearing oven mitts.

Once you've navigated the labyrinthine menu and set your desired temperature, the Cybertruck takes this as a mere suggestion rather than an actual command. Say you want a cozy 21 degrees Celsius. The truck acknowledges this, then appears to consult its own mood before deciding whether you'll be basking in tropical heat or shivering in Arctic conditions.

The system's idea of 21 degrees is so subjective, it could be part of an abstract art installation titled 'The Concept of Warmth'. One day, 21 degrees will have you opening windows to let out the heat, pondering if the truck is trying to double as a sauna. The next day, the same setting will have you wrapped up like an Arctic explorer, seeing your breath mist up in the air, wondering if you've somehow been transported to Siberia.

And then there's the fan speed, which appears to have only two settings: 'hurricane' and 'gentle summer breeze'. The middle ground is a myth, a legend, like the Loch Ness Monster or a quiet Lamborghini. On the hurricane setting, the fan noise drowns out all conversation, making communication in the truck resemble a scene from a silent movie.

The placement of the air vents in the Cybertruck is another exercise in Tesla's unique brand of creativity. They're so well hidden that finding them is like going on a treasure hunt, but instead of treasure, you're searching for the source of the gale-force wind or the faint puff of air that's supposedly heating or cooling the cabin.

Let's not forget about the automatic climate control feature, which is about as reliable as a chocolate fireguard. The truck takes into account factors like the outside temperature, the position of the sun, and probably the alignment of the planets, to decide what it thinks the cabin temperature should be. The result is a climate that changes so frequently, you start to feel like you're living in a time-lapse video of different weather patterns.

Adjusting the climate control while driving is a task that requires the multitasking abilities of an air traffic controller. One eye on the road, one eye on the gigantic touchscreen, one hand on the wheel, and one hand attempting to adjust the temperature without accidentally opening the trunk or engaging autopilot.

In conclusion, the climate control system in the Cybertruck is a masterpiece of unpredictability. It's a system that takes your input, laughs in the face of logic, and then does whatever it pleases. You don't control the climate in the Cybertruck; you merely suggest and hope. It's part of the adventure, the thrill, the unique experience of owning a piece of the future, even if that future seems to have been programmed by a team of whimsical elves who've never experienced weather. So buckle up, set your temperature, and enjoy the ride – it's going to be a wild one, climatically speaking.

## The Sauna vs. Arctic Tundra Setting

In the realm of the Tesla Cybertruck, the climate control system doesn't believe in middle grounds. It operates in absolutes – you're either recreating the conditions of a sauna in the depths of Finland, or you're emulating the chill of an Arctic tundra. This isn't climate control; it's climate roulette.

Firstly, let's address the Sauna setting. You would think setting the temperature to a mild 22 degrees Celsius would result in a comfortable, balmy environment. But no, the Cybertruck interprets 22 degrees as 220. The heat erupts from the vents with the subtlety of a volcano. Within minutes, you're not driving; you're marinating in your own sweat, peering through a mist of humidity that would make a tropical rainforest seem arid. Conversations in the car quickly turn to discussions about the weather, global warming, or how it feels to be a roast chicken.

Attempting to adjust the temperature results in a series of frantic stabs at the touchscreen, which is more sensitive than a poet on Valentine's Day. Each touch is a gamble. Will it lower the temperature, or will it interpret your command as a request to increase the intensity of your impromptu steam bath?

Then we have the Arctic Tundra setting. This is the Cybertruck's interpretation of 'cooling down'. What you want is a gentle, refreshing breeze to waft through the cabin. What you get, however, could freeze the balls off a brass monkey. The air conditioning kicks in with the ferocity of a blizzard. You can almost see the icicles forming on the rearview mirror. Passengers start rummaging for winter coats and considering the merits of a hot chocolate dispenser in the glove compartment.

The disparity between these two settings is like choosing between being slapped or tickled – both are startling in their own way, and neither is particularly comfortable. The middle setting, which in any sensible car would provide a nice, moderate temperature, is a mythical land in the Cybertruck. It's like finding a unicorn. You've heard stories about it, you hope it exists, but you've never actually seen it.

Then there's the matter of trying to change these settings while driving. It's like trying to solve a Rubik's cube in a rollercoaster. You're bouncing along, trying to keep one eye on the road while the other navigates through a maze of menus and submenus, desperately trying to find the elusive 'just right' temperature. It's a test of dexterity, focus, and patience – a triathlon for your fingers and your sanity.

The Cybertruck's climate control system is less a feature and more a conversation starter. It's like having a moody pet – you never know what you're going to get. Sometimes it behaves, providing a comfortable environment as you cruise along. Other times, it throws a tantrum, blasting you with hot or cold air as if to say, "I control the weather in here, human."

In conclusion, mastering the climate control in the Cybertruck is akin to taming a wild beast. It requires patience, a gentle touch, and perhaps a small sacrifice to the gods of technology. Whether you're sweltering in the Sauna setting or shivering in the Arctic Tundra mode, remember – it's all part of the Cybertruck experience. It's not just a drive; it's an endurance test, a battle of wills between man and machine. So next time you get in your Cybertruck, ask yourself – is it hot in here, or is it just me? The answer will always be a resounding, emphatic "Yes."

## Guess which vent will actually blow air today!

In the Cybertruck, the climate control system is a bit like playing a game of Russian roulette, but with vents. Every day is a new adventure in automotive climate unpredictability, a thrilling mystery of "Guess Which Vent Will Actually Blow Air Today!" It's Tesla's way of adding a bit of spice to the mundane task of air circulation, turning your daily commute into a game show hosted by the truck itself.

Let's set the scene. You enter the Cybertruck, a vehicle that looks like it was designed by someone who thought Blade Runner wasn't futuristic enough. You expect cutting-edge technology, but what you get with the vent system is more akin to playing a slot machine – you pull the lever and hope for the best.

When you press the button to activate the climate system, you expect a harmonious symphony of vents working together to bring you comfort. Instead, it's like each vent is an independent entity, a free spirit unwilling to conform to the norms of vent behavior. One vent decides it's a tropical day in the Bahamas, blowing warm air with the enthusiasm of a holidaymaker.

Another vent opts for a more Arctic approach, providing a stream of air so cold, you wouldn't be surprised to see penguins marching out of it.

The fun really begins when you try to adjust the direction of these vents. In a regular car, you have physical vents that you can point in whichever direction you like. In the Cybertruck, it's all controlled via the touchscreen – because physical dials are so last century. Adjusting the airflow direction is like trying to conduct an orchestra with a broken baton. You swipe, you tap, and what happens? The vent you wanted to adjust remains stoically still, while the one on the other side of the car suddenly springs to life, blowing papers and light objects around the cabin like a mini tornado.

And then there's the mystery of the disappearing airflow. You're driving along, the cabin finally at a comfortable temperature, when suddenly one vent decides it's time for a nap. The airflow stops without warning, as if the vent has been placed in witness protection and vanished without a trace. You prod at the touchscreen, but the vent has made up its mind – it's on strike.

The center vents are particularly special. They seem to have a mind of their own, sometimes operating as intended, other times acting as nothing more than decorative pieces, as useful as a chocolate fireplace. You can almost hear them laughing at your futile attempts to coax some air out of them.

And let's not forget the excitement of the defrost feature. You activate it expecting a clear windshield, but the Cybertruck interprets this as a suggestion rather than a command. The windshield may clear, or you may just get a nice, even fogging, as if the truck is trying to recreate a romantic, misty scene from a film noir.

In conclusion, the vent system in the Cybertruck is less about climate control and more about climate chaos. It's a daily gamble, a spin of the roulette wheel where the prize is either a blast of Arctic chill or a waft of desert heat. Each journey in the Cybertruck becomes a foray into the unknown realms of HVAC systems. It's not just a drive; it's a saga, an epic tale of man versus machine, where the quest is for a consistent stream of air. So buckle up, choose your settings, and enjoy the game of Vent Roulette. Just remember, in the Cybertruck, the house always wins.

## Seat "Adjustments" via Esoteric Levers

In the bizarre and often incomprehensible world of the Tesla Cybertruck, the act of adjusting your seat is transformed from a mundane task into a cryptic ritual, akin to deciphering the Dead Sea Scrolls. The concept of a simple lever to adjust your seat has been tossed out of the window, along with common sense and any semblance of user-friendliness.

Upon entering the Cybertruck, one is immediately confronted with what can only be described as an enigma in the form of a seat. Tesla, in their infinite wisdom, has replaced the traditional, understandable seat adjustment levers with a series of controls that resemble something from the cockpit of an alien spacecraft. These esoteric levers are about as intuitive as reading hieroglyphics without a Rosetta Stone.

Attempting to adjust the seat for the first time is like trying to crack the Da Vinci Code. You're greeted with a series of levers, buttons, and sliders, each more mysterious than the last. The levers, adorned with cryptic symbols, seem to follow no logical order or ergonomics. You pull one lever, expecting the seat to move back, and suddenly you're in a fully reclined position, staring at the ceiling of the truck as if you're about to receive dental surgery.

Then there's the lever for lumbar support, a feature that should be straightforward but is instead shrouded in mystery and confusion. You adjust it, hoping for some relief for your aching back. However, the Cybertruck interprets this as an invitation to either aggressively push your spine into a contorted position or withdraw all support, leaving you slumped like a deflated balloon.

Let's not forget the height adjustment lever. In any normal vehicle, this would simply raise or lower your seat. But in the Cybertruck, it's a gamble. Pull the lever, and you might ascend gracefully, getting a better view of the road. Or you might find yourself sinking slowly, as if the seat is swallowing you whole, leaving you with the driving position of a toddler behind the wheel.

The seat adjustment levers are not just controls; they're a puzzle, a test of your problem-solving skills, and a lesson in patience. Each adjustment is a foray into the unknown. Will you find a comfortable driving position, or will you end up contorted like a circus performer, limbs akimbo, with a crick in your neck and a twitch in your eye?

And it's not just the driver's seat. Oh no, Tesla wouldn't want the passengers to feel left out of this fun game. Each passenger seat comes with its own set of baffling levers, ensuring that everyone in the truck can partake in the joyous activity of trying to find a comfortable seating position. It's like a family board game, but less fun and more likely to end in a trip to the chiropractor.

In conclusion, adjusting your seat in the Cybertruck is not a simple task; it's an adventure, a journey through a maze of levers and buttons, each more puzzling than the last. It's a daily challenge, a test of your dexterity, your intuition, and your ability to contort your body into positions you never thought possible. So, buckle up, grab the levers, and prepare for a ride. Just remember, in the Cybertruck, comfort is not a destination; it's a journey. A long, bewildering, and slightly uncomfortable journey.

## Pull a lever and see what moves - if anything.

In the beguiling world of the Tesla Cybertruck, every interaction with the interior feels like participating in a lever lottery. Each lever is a mystery, a gamble, a game of mechanical Russian roulette where you're never quite sure what will happen. Will pulling a lever adjust your seat, or will it eject you through the roof like a startled astronaut? Only one way to find out!

Let's start with what I can only describe as the "seat adjustment" levers, although labeling them as such might be giving them too much credit. In any ordinary vehicle, pulling a lever under the seat prompts a predictable and boring response: the seat moves forward or backward. But in the Cybertruck, such mundane predictability is scoffed at. Here, pulling a lever is an invitation to a surprise party hosted by the seat itself. Will the seat slide gracefully back? Will it abruptly drop you into a chiropractic nightmare? The suspense is unbearable.

One lever, in particular, seems to be connected to nothing and everything at the same time. Pull it, and sometimes the backrest tilts, offering a less-than-gentle massage to your spine. Other times, it appears to do absolutely nothing, leaving you yanking at it like a frustrated gambler at a broken slot machine. And on rare, magical occasions, it might just adjust the seat as intended – but don't hold your breath.

Then there's the lever lottery for the steering column. Adjusting the steering wheel should be a simple, straightforward affair. Not so in the Cybertruck. Here, you pull a lever, and rather than the wheel moving up or down, it either retracts so far into the dashboard that you might as well steer with your teeth, or it comes so close that it's practically giving you a hug. Finding the sweet spot requires the precision of a bomb disposal expert.

The height adjustment lever is where things get really interesting. One would assume that pulling it would raise or lower your seat. But assumptions are for the mundane, and the Cybertruck is anything but. Pull this lever and brace yourself. Will the seat ascend like a throne, granting you a commanding view of the road? Or will it drop like a stone, leaving you peering over the dashboard like a child on a booster seat? The excitement is palpable.

And let's not forget the lever that adjusts the lumbar support, which in the Cybertruck is less 'support' and more 'guesswork'. You pull it expecting some relief for your lower back, and suddenly you're being pushed forward, forced into a driving position that suggests you're about to enter a high-speed chase. Alternatively, it might just vanish entirely, offering all the support of a politician's promise.

Operating these levers isn't just an action; it's an experience, a journey into the unknown. Each pull is a leap of faith, a trust exercise between man and machine. You enter the Cybertruck not just as a driver, but as an explorer, charting a course through the uncharted territory of Tesla's interior design philosophy.

In conclusion, the levers in the Cybertruck provide a daily dose of excitement and uncertainty. Every journey is accompanied by the thrill of the unknown – a lever lottery where the stakes are your comfort and sanity. It's a game, a challenge, a test of your intuition and your willingness to adapt. So go ahead, pull a lever, and embrace the adventure. Just be prepared for the unexpected – because in the Cybertruck, that's the only thing you can count on.

## Comfort? More Like a Torture Device

In the grand and perplexing design of the Tesla Cybertruck, one feature stands out for its sheer audacity to redefine discomfort – the seats. You see, Tesla, in its pursuit of futuristic minimalism, appears to have modeled the Cybertruck's seats not on something you sit in, but rather on medieval torture devices. The sort of thing you'd find in the Tower of London, not in a vehicle.

Sitting in these seats, you quickly realize that 'comfort' was probably at the bottom of the priority list, somewhere below 'make sure it looks like it was designed by a hyperactive origami enthusiast'. The seats, which Tesla might claim are ergonomically designed, seem to have been ergonomically designed for a species yet to be discovered. For the human anatomy, they are about as comfortable as a bed of nails.

The lumbar support, or what Tesla optimistically refers to as support, is a misnomer. It's more akin to a lumbar assault. The moment you lean back, you're greeted with a protrusion that feels like a fist strategically positioned to remind your spine of every life choice it's ever regretted. Adjusting this so-called support doesn't alleviate the discomfort; it merely shifts it to another part of your back, in a malevolent game of whack-a-mole.

Then there's the seat's cushioning, which seems to have been inspired by the stone slabs used by medieval monks. It's as if Tesla decided that padding was an unnecessary luxury, an obsolete relic from a bygone era of automotive design. Within minutes of sitting, you start to wonder if you've accidentally signed up for an endurance test, a trial of your resolve and pain threshold.

The headrest, too, deserves a special mention. Far from providing a rest for your head, it feels more like a device designed to prevent any form of relaxation. It juts forward, ensuring your neck is at an angle that chiropractors dream of, the sort of position that screams 'future business'.

And let's not forget the width of the seats. Designed for what can only be described as a prototype human, the seats are so

## Mood Lighting That Reflects Your Disappointment

In the realm of the Tesla Cybertruck, where innovation meets confusion, the concept of mood lighting takes on a whole new dimension. It's like Tesla decided that what drivers really need is lighting that not only reflects but also amplifies their disappointment. This isn't just mood lighting; it's moodiness lighting.

As you step into the Cybertruck, you're greeted by an array of lights that seem to have been strategically placed to highlight the vehicle's stark interior. The lighting is less about creating an ambiance and more about ensuring you can fully appreciate the lack of comfort and ergonomic design. It's like shining a spotlight on a plate of overcooked vegetables – it doesn't make it any more appetizing.

The choice of colors for the mood lighting is, well, moodier than a teenager who's just been told they can't go to a party. You have options, but they range from 'depressing blue' to 'existential crisis grey'. There's no warm, welcoming yellow or cheerful green. No, Tesla's color palette for mood lighting is seemingly inspired by a dystopian future where joy has been outlawed.

Then there's the control for the mood lighting, buried in the depths of the infotainment system, because why make anything easy? Adjusting the lights requires the precision of a surgeon and the patience of a saint. You navigate through menu after menu, each click a gamble. Will you find the lighting controls, or will you accidentally trigger the ejection seat? With the Cybertruck, you can never be too sure.

Once you've finally found the controls and adjusted the lighting to your liking – or as close to your liking as the options will allow – you settle into the drive. But then, the lights seem to have a life of their own. They dim and brighten at random intervals, independent of any input from you. It's like they're responding to your thoughts, your mood, your very soul – and they've decided that what your soul really needs is an inconsistent and mildly infuriating light show.

The mood lighting's piece de resistance is its reaction to music. Play something upbeat, and the lights don't dance or pulse in time with the music. Oh no, that would be too pleasurable. Instead, they remain steadfastly, stubbornly static, a visual representation of Tesla's commitment to sucking the joy out of every situation.

And let's not forget when the mood lighting decides to malfunction. Flickering like a candle in a windstorm, it doesn't so much create an atmosphere as it does a sense of impending doom. It's like being in a horror movie, right at the moment before the monster jumps out. Except there's no monster, just the creeping realization that you've spent a small fortune on a truck that can't even get mood lighting right.

In conclusion, the mood lighting in the Cybertruck is a marvel of modern technology in the same way that a rainstorm is marvelous to someone without an umbrella. It's less about enhancing the driving experience and more about reminding you that no matter how advanced technology gets, it can still be utterly, incomprehensibly disappointing. So sit back, bask in the glow of your dismal lighting, and remember: this is the future, apparently.

### Discover the lighting that barely illuminates but maximizes despair.

In the cavernous, angular world of the Tesla Cybertruck, the interior lighting is less about illumination and more about fostering a dim glow of regret. It's as if the designers, in a moment of bleak introspection, decided that what drivers really needed was a lighting system that barely allows you to see, but maximizes your sense of despair.

Upon entering the Cybertruck, you're greeted not by a warm, welcoming glow, but by a series of lights that seem to have been designed by someone who finds the concept of happiness offensive. The lighting is so minimal, it makes a candlelit dinner look like a police interrogation room. It's like Tesla took the idea of mood lighting, sucked all the joy out of it, and then dimmed it down just for good measure.

The main cabin lights, which in any normal vehicle would adequately light the interior, emit what can only be described as the automotive equivalent of a sigh. It's enough light to make out shapes and forms, but not enough to read a map, a book, or the expression of mounting frustration on your passenger's face.

Then there's the dashboard lighting, a crucial aspect in any car. But in the Cybertruck, Tesla has decided that what you really need is a dashboard that's lit like a failed nightclub. The instruments are bathed in a sort of gloomy glow that makes you squint and guess at your speed, battery level, and whether that warning light is serious or just the truck's way of saying, "I'm not happy."

The touchscreen, the centerpiece of the Cybertruck's interior, is a beacon of despair in the dimly lit cabin. It's bright enough to ensure it attracts all your attention but not bright enough to be particularly useful. It's like a lighthouse that's running on half power – you can see it, but it's not really helping you navigate.

Tesla, in their infinite wisdom, included ambient lighting in the Cybertruck. But rather than creating an ambient, relaxing environment, it's more like sitting in a bar that's trying too hard to be cool. The ambient lighting doesn't so much illuminate as it does cast shadows, creating a gloomy atmosphere that would make even a goth think, "Maybe this is a bit too much."

Adjusting the lighting in the Cybertruck isn't a task; it's a journey – a journey into the depths of a menu system that's as user-friendly as a Rubik's Cube. You tap, swipe, and prod, trying to find a setting that's bright enough to see, but the truck seems to take a perverse pleasure in keeping you in the dark, both literally and metaphorically.

And when you finally do find a setting that provides more light, it's a harsh, unforgiving glare that makes you yearn for the dimness you had before. It's like choosing between being in the dark and being interrogated under a spotlight. Neither is particularly pleasant, but at least the darkness was more forgiving.

In conclusion, the lighting in the Tesla Cybertruck is an exercise in minimalism taken to a depressive extreme. It's as if the truck is trying to make a statement, not about efficiency or design, but about the futility of existence. The dim glow of regret that permeates the cabin is a constant reminder that no matter how advanced we become, no matter how futuristic our vehicles are, we're still just people sitting in the dark, trying to find the light switch. So sit back, enjoy the gloom, and remember – in the Cybertruck, even the light at the end of the tunnel is probably just an oncoming train.

## When the lights decide to have a rave without your consent.

In the seemingly boundless oddity that is the Tesla Cybertruck, the lighting system occasionally takes on a life of its own, throwing an impromptu rave without your consent. It's as if the lights, bored of their monotonous existence, decide to throw a party, turning your serene drive into an unwanted disco.

Imagine you're cruising along, enjoying the monastic silence that only an electric vehicle can provide. Suddenly, without warning, the interior lighting begins to flicker like a strobe light at a 90s rave. It's disconcerting, disorienting, and gives you a fleeting sense of empathy for a disco ball. One moment you're in serene darkness, the next you're part of an unsolicited light show.

This isn't just any flickering – it's as if the lights are possessed by the spirit of a retired DJ who's decided to have one last hurrah. The flickering doesn't follow any discernible rhythm or pattern; it's erratic, unpredictable, and has more mood swings than a teenager.

You fumble for the controls, jabbing at the touchscreen in a futile attempt to restore order. But the touchscreen, being the touchscreen, decides that this is the perfect moment to freeze, leaving you helplessly watching as your peaceful drive turns into a scene from a low-budget science fiction film.

The flickering isn't limited to the interior lights. Oh no, that would be too kind. The headlights join in, turning your journey into a beacon for confused club-goers. To the outside world, your Cybertruck isn't just a vehicle; it's a mobile nightclub, complete with its own light show. Pedestrians stop and stare, half-expecting a door to open and a bouncer to step out, asking if they're on the list.

And then there's the effect on your driving. Trying to concentrate on the road while your car is throwing an impromptu party is like trying to read a book in a washing machine. The flickering lights create a kaleidoscope of shadows, transforming the mundane into the surreal. Every tree, every sign, every passerby is cast in an eerie, pulsating glow.

Let's not forget how the flickering plays with your mind. At first, it's merely annoying. Then it becomes a test of your sanity. You start to wonder if you're imagining it, if perhaps you've accidentally ingested some hallucinogenic substance. You glance at your passenger, searching for a sign of shared experience, a confirmation that you haven't slipped into an alternate reality.

The cause of this flickering fiesta? It could be anything. A software glitch, a loose wire, or perhaps the Cybertruck has developed a consciousness and is expressing its newfound creativity. Tesla might call it a 'feature', an 'unexpected easter egg', but you know the truth. It's not a feature; it's a bug, a gremlin in the system having a laugh at your expense.

In conclusion, when the lights in your Cybertruck decide to have a rave without your consent, you're not just flickering into the abyss; you're experiencing a unique feature of Tesla ownership. It's a reminder that in a world of cutting-edge technology, sometimes the cutting edge can be a bit too sharp. So sit back, try to relax, and enjoy the light show. After all, who needs a nightclub when you've got a Cybertruck? Just keep your eyes on the road, your hands on the wheel, and maybe, just maybe, carry a torch.

# 5. Entertainment or Lack Thereof

When you clamber into the Tesla Cybertruck, expecting a cornucopia of entertainment options akin to a Las Vegas casino, you're in for a shock. The entertainment system in this behemoth is more akin to a Victorian-era parlor room – you're aware that entertainment is supposed to happen here, but you're not quite sure how.

Firstly, let's talk about the infotainment screen, a behemoth in itself. This screen is so large it could double as a dining table. You'd think with all this real estate, the entertainment options would be endless. But no, Tesla, in a fit of what can only be described as madness, decided that simplicity is the key. So simple, in fact, that figuring out how to play your favorite tunes becomes a task as complex as diffusing a bomb. The interface looks like it was designed by someone who thought that a Rubik's Cube wasn't puzzling enough.

Then there's the music system. The Cybertruck comes equipped with speakers that, at first glance, promise an auditory experience akin to a live concert. However, when you start playing music, it quickly becomes apparent that the sound quality is more garage band than Royal Albert Hall. It's like listening to Beethoven being played through a tin can connected to a string.

And should you wish to listen to the radio – an archaic concept in Tesla's world – you'll find the experience similar to trying to tune a radio in the 1920s. It's a hissy, crackly affair, where finding a station that doesn't sound like it's being broadcast from the moon is a cause for celebration.

Let's not forget the navigation system's attempts at being part of the entertainment package. It seems to think that taking you on the scenic route, no matter your destination, is a form of amusement. You ask it to take you to the nearest supermarket, and suddenly you're on a magical mystery tour through the countryside. It's not just navigation; it's an adventure – one you didn't sign up for.

The voice control system, which should be a highlight, has all the functionality of a parrot with a limited vocabulary. You ask it to play The Beatles, and it either doesn't understand, or in a fit of rebelliousness, plays The Rolling Stones instead. It's like having a surly teenager as your personal DJ.

And for those in the back seats, the entertainment is even more sparse. There are no screens, no gadgets, nothing but the joy of looking out at the world passing by. It harks back to a simpler time, a time when a stick and a hoop were considered the height of entertainment. It's character-building, according to This manual. In reality, it's boredom-inducing.

In conclusion, the entertainment system in the Tesla Cybertruck is a paradox. It promises so much yet delivers so little. It's a system that confounds, frustrates, and occasionally amuses – but not in the way it was intended to. It's like being given a book with all the pages blank and being told it's a novel. The entertainment in the Cybertruck is an exercise in minimalism, a test of your ability to entertain yourself. So, buckle up, enjoy the silence, and if you want entertainment, I suggest bringing a book. Or a hoop and a stick.

## The AM Radio Experience: Static and All

In the futuristic hulk of the Tesla Cybertruck, where everything is supposed to be as cutting-edge as a freshly sharpened guillotine, the AM radio experience is akin to stepping back into a bygone era. It's like Tesla, in their relentless pursuit of innovation, accidentally stumbled into a time warp and brought back an AM radio from the 1950s. The only thing missing is a crackling fireside chat.

Turning on the AM radio in the Cybertruck is an experience that can only be described as auditory time travel. As the first crackles and hisses of static hit your ears, you're instantly transported to a time when radio was the height of home entertainment, and television was just a fad. The static isn't just a feature; it's an entire character in the Cybertruck's ensemble of quirks.

As you attempt to tune into a station, the Cybertruck's AM radio responds with the enthusiasm of a sleepy cat. The dial moves reluctantly, begrudgingly, as if it's not quite ready to let go of the ghost of radio past. Finding a station amidst the sea of static is like trying to find a needle in a haystack, blindfolded, with one arm tied behind your back.

And when you do manage to latch onto a station, it's never quite clear. It's a symphony of crackles, pops, and the faint murmur of human voices. It's like listening to a conversation in another room, through a wall, during a hailstorm. You strain your ears, trying to piece together words and sentences, but it's like your grandma trying to use Snapchat – confusing and slightly surreal.

The AM radio in the Cybertruck doesn't just play music or news; it plays a guessing game. Is that a song, or is the static just forming a rhythmic pattern? Is the announcer discussing the weather, or is it a recipe for apple crumble? The answers are elusive, lost in the white noise that fills the cabin.

And let's talk about the joy of interference. Driving under a bridge? Static. Passing a tall building? More static. A cloud looks at you funny? That's right, even more static. The AM radio in the Cybertruck is more sensitive to interference than a politician to a difficult question. It's as if the radio is playing a game, and the only rule is that there are no rules – only static.

But it's not all bad. Oh no, the AM radio in the Cybertruck provides a sense of nostalgia, a throwback to simpler times when radio was magic and the world was a bigger, more mysterious place. It's an escape from the cold, clinical digital world we live in, a reminder that sometimes, the old ways had a charm and a warmth that modern technology can't replicate.

In conclusion, the AM radio experience in the Tesla Cybertruck is an anachronism, a delightful and infuriating throwback to the golden age of radio. It's a mix of frustration, nostalgia, and a healthy dose of bewilderment. So next time you're in your Cybertruck, turn on the AM radio, sit back, and enjoy the static. It's not just noise; it's a history lesson. A reminder of a time when listening to the radio required patience, imagination, and a tolerance for a never-ending symphony of crackles and hisses. Welcome to the past, courtesy of the future.

How to find that one station that's not just white noise.

In the realm of the Cybertruck, where the future is now and the past is a dim memory, the AM radio stands as a defiant relic. It's like stumbling upon a cave painting in the age of virtual reality. Tuning into a station amidst the relentless white noise of the AM radio in the Cybertruck is akin to Indiana Jones searching for the Lost Ark – it requires determination, a bit of luck, and a sense of wonder at the ancient technology.

As you embark on this auditory archeological dig, the first thing you realize is that the tuning process is less science and more witchcraft. You turn the dial, a process that should be straightforward, but here in the Cybertruck, it's a journey through static, a symphony of crackles, hisses, and the occasional beep. It's like trying to communicate with extraterrestrial life using a toaster.

As you navigate through the sea of white noise, you start to wonder if there are any stations broadcasting at all, or if Tesla, in a fit of postmodern irony, has designed the radio to only play static as a commentary on the state of modern media. But then, faintly at first, you hear it – the murmur of human voices, the hint of a melody. It's a radio station, broadcasting from the distant land of 'Not Just Static'.

Finding this station is a moment of triumph, akin to summiting Everest or baking the perfect soufflé. You sit back, a smug smile on your face, ready to bask in the warm glow of your success. But this is the Cybertruck, and nothing is that easy. The moment you settle in, the signal begins to fade, the voices and music drowning in a growing tide of static. It's like the truck is teasing you, offering a glimpse of what could be, only to snatch it away with a crackle and a pop.

And so, the battle continues. You adjust the dial, millimeter by millimeter, fighting for every scrap of signal like a starving seagull on a beach. Each tiny movement brings a change – a little less static here, a little more voice there. It's a delicate dance, a balancing act between hope and despair.

But let's say you do it. You find that sweet spot, where the static recedes into the background, and the voices and music come through clear and strong. It's a miraculous moment, a break in the clouds on a stormy day. You've done it. You've tuned into the past and found that one station that's not just white noise.

Listening to this station, you're transported back in time. The music, the news, the entire ambiance of the broadcast is a reminder of days gone by, a time capsule preserved in the airwaves. It's nostalgic, it's comforting, it's... fading again. Yes, the signal is on the move, and you're back in the fight, chasing it around the dial like a cat after a laser pointer.

In conclusion, finding a station on the Cybertruck's AM radio that's not just white noise is an adventure, a challenge, a test of your will and patience. It's a throwback to a simpler time, a reminder that not so long ago, entertainment required effort and engagement. So when you sit in your Cybertruck, remember: tuning into the past is not just about finding a radio station. It's about rediscovering the joy of the hunt, the thrill of the chase, and the sweet satisfaction of snatching victory from the jaws of relentless, never-ending static. Welcome to the past, courtesy of the future – crackles, hisses, and all.

## Embracing the static as part of your musical journey.

In the bizarre world of the Tesla Cybertruck, where everything is supposed to be futuristic and cutting-edge, the AM radio stands as a defiant throwback to the days when listening to the radio was more about battling interference than actually enjoying music. In this vehicle, static isn't just part of the listening experience; it is the experience.

Now, let's be clear. When you first attempt to listen to the radio in the Cybertruck, you're expecting crystal clear sound, an auditory experience so sublime that angels weep. What you get instead is a symphony of crackles, hisses, and pops that sound like they're being transmitted from the far side of the moon. But here's where we embrace the madness. This isn't just interference; it's ambient, avant-garde music.

As you embark on your drive, the static becomes a constant companion. It's like having a slightly annoying friend who talks too much – you can't get rid of them, so you might as well make the best of it. Every crackle and hiss starts to take on a personality. That long whine as you drive under a power line? That's just the radio saying hello. The crackling that sounds like bacon frying? That's the sound of the radio getting comfortable.

Trying to find a station amidst this cacophony is akin to panning for gold. You know there's something valuable in there, but you have to sift through a lot of silt to find it. Every now and then, you catch a snippet of a song, a ghostly fragment of a melody that fades into the static as quickly as it appeared. It's like catching a glimpse of a unicorn in a forest – magical, but infuriatingly elusive.

But let's not despair. This is where we start to find joy in the interference. You see, unlike the sterile perfection of digital radio, AM radio with its interference is alive. It's organic. It changes with the landscape, the weather, even your speed. It's interactive in a way that modern technology can't replicate. You're not just listening to the radio; you're participating in a live performance.

And there's something oddly calming about the constant wash of static. In a world that's always connected, always clear, always perfect, the imperfection of AM radio is a balm for the soul. It's a reminder that not everything has to be crystal clear to be enjoyable. Sometimes, the noise, the mess, the chaos is part of the fun.

As you drive, the static and the occasional fragments of music become the soundtrack to your journey. You start to find rhythms in the noise, patterns in the chaos. It's like listening to the world's most avant-garde jazz musician playing a solo just for you. And when you do finally tune into a station, the music sounds all the sweeter for the effort it took to find it.

In conclusion, the AM radio in the Tesla Cybertruck, with its joy of interference, is a lesson in embracing imperfection. It's a throwback to a simpler time, a nod to the days when listening to the radio was an adventure, not just a background noise. So next time you're in your Cybertruck, turn on the AM radio, turn up the static, and enjoy the ride. It's not just a journey; it's a musical adventure, complete with its own unique soundtrack of snaps, crackles, and pops. Welcome to the joy of interference – the most unexpectedly authentic musical journey you'll ever take.

## Pairing Your Device with the Infuriotainment System

In the world of the Tesla Cybertruck, where the future is apparently now, and common sense was left at a charging station miles back, pairing your device with the infotainment system is an exercise in frustration and bewilderment. It's like trying to teach a cat to fetch – theoretically possible, but in reality, a journey into madness.

First off, the term 'infotainment' in the context of the Cybertruck is what I'd call a bit of a misnomer. 'Infuriotainment' might be more appropriate because, by the end of the process, you're not entertained; you're infuriated. You begin this technological odyssey with high hopes. "Surely," you think, "in such a futuristic vehicle, pairing my device will be a breeze." Oh, how naïve.

The process starts innocuously enough. You find the Bluetooth settings in the menu that's only slightly less complicated than a nuclear submarine control panel. The car is now in 'pairing mode', a state that's about as welcoming as a doorman at an exclusive club. Your device, sensing the impending struggle, is already quivering in your hand.

As you attempt to pair your device, the Cybertruck's infotainment system seems to take a moment to consider your request, like a cat deciding whether to acknowledge your existence. Moments pass. The tension builds. And then, it rejects your pairing attempt with the disdain of a French waiter turning away a tourist in shorts.

You try again. This time, the system acknowledges your device, but just when you think you're about to succeed, it asks for a PIN – a series of numbers that you're pretty sure was never mentioned in This manual. It's a guessing game, and the Cybertruck holds all the answers.

Suppose you manage, through a combination of sheer luck and blind persistence, to guess the PIN. You're in. But victory is short-lived. Now, your device is connected, but there's no sound. The infotainment system displays the track playing, but the sound is lost in the ether – perhaps it's entertaining the electrons in the vehicle's electrical system.

So, you delve back into the menu, navigating through settings that seem to have been designed by someone who took inspiration from a labyrinth. You tweak, you adjust, you mutter increasingly colorful language under your breath. And then, out of nowhere, sound erupts from the speakers. It's loud, it's unexpected, and it's the wrong song.

You scramble to adjust the volume, but in your panic, you activate the voice control, which is now asking you, in a tone that suggests it's judging you, what assistance you require. You stammer a response, but it doesn't understand. Why would it? You're now speaking a language comprised entirely of panic and confusion.

By now, a simple drive has turned into a saga. You've battled with Bluetooth, wrestled with a PIN, lost a fight with the volume, and been condescended to by a voice control system that's about as helpful as a chocolate fireguard.

In conclusion, pairing your device with the Cybertruck's infotainment system is not for the faint of heart. It's a test of will, a battle of human versus machine, a modern-day David and Goliath story, where Goliath is a car stereo, and David is left wondering why he didn't just bring a CD. But once you have paired your device, you'll feel a sense of accomplishment akin to summiting Everest or baking the perfect soufflé. Just remember, in the Cybertruck, the journey to musical nirvana is a long and arduous one, paved with confusion, frustration, and a voice control system that, quite frankly, could use some therapy.

## Bluetooth or Blue-rage?

In the realm of the Tesla Cybertruck, attempting to connect your phone via Bluetooth is an experience akin to trying to solve a Rubik's cube while blindfolded and riding a unicycle. The process, which in any other vehicle would be as simple and mundane as making toast, in the Cybertruck is transformed into an epic saga of frustration, confusion, and eventual blue-rage.

First, you embark on what should be a straightforward task: finding the Bluetooth settings in the infotainment system. But in the Cybertruck, this is the first trial. The menu is less a user-friendly interface and more a labyrinth designed by a sadist. You click through screens that seem to have been arranged by someone playing Pin the Tail on the Donkey. After several minutes of increasingly frantic tapping, you find the Bluetooth menu, hidden away like a treasure in a pirate novel.

With naive optimism, you select 'Pair new device', and your phone begins the search. The Cybertruck, meanwhile, appears to enter a state of deep contemplation, as if pondering the meaning of life rather than performing a basic function. Minutes tick by. Your phone, ever hopeful, keeps searching, while the Cybertruck's screen displays a spinning wheel of doom, mocking your futile attempts at connectivity.

Then, miraculously, your phone finds the Cybertruck. But this is not the end; it's merely the end of the beginning. You select the Cybertruck on your phone, and a PIN appears on the vehicle's screen, a series of numbers that you're convinced is the solution to some ancient mathematical problem. You enter the PIN into your phone, and for a brief, shining moment, you believe you've succeeded.

But no. The Cybertruck rejects your phone like an immune system fighting off a particularly aggressive virus. You try again, re-entering the PIN with the precision of a bomb disposal expert. Yet, the result is the same: connection failed. The screen on the Cybertruck remains as impassive as the Queen's Guard.

Undeterred, you forge ahead, rebooting your phone, turning the vehicle's Bluetooth off and on, and performing the technological equivalent of standing on one leg and facing north. You repeat the pairing process, each attempt chipping away at your sanity like a sculptor creating a masterpiece of despair.

And then, when all hope seems lost, when you're on the brink of hurling your phone out of the window and communicating via carrier pigeon, it happens. The phone and the Cybertruck connect. A small notification appears, as underwhelming as finding a single biscuit left in the tin. You've done it. You've conquered the beast. Your phone is connected. Music, contacts, and all the wonders of your device are now at your fingertips, integrated into the Cybertruck's system.

But the victory is bittersweet. The process has taken years off your life, turned your hair a little grayer, and introduced you to new levels of frustration. You realize that connecting your phone to the Cybertruck isn't a task; it's a quest, a Herculean trial, a test of endurance, patience, and willpower.

In conclusion, connecting your phone to the Cybertruck via Bluetooth is an experience that oscillates wildly between hope and despair. It's a journey, a saga, a narrative of man versus machine where, against all odds, man occasionally triumphs. But the price of victory is steep, paid in the currency of sanity and time. So, the next time you step into your Cybertruck and consider connecting your phone, ask yourself: is it worth it? And then, when you inevitably decide that it is, brace yourself for the epic tale of Bluetooth or Blue-rage. Remember, in the world of the Cybertruck, nothing is easy, but everything is memorable.

## The Eternal "Loading" Screen: Patience is a virtue

In the futuristic yet seemingly backward world of the Tesla Cybertruck, the infotainment system introduces you to a new level of patience-testing: the eternal "Loading" screen. This isn't just a minor inconvenience; it's a lesson in the art of Zen, a test of endurance where your playlist is perpetually three tantalizing seconds away from blessing your ears.

Let's set the scene. You're comfortably ensconced in your Cybertruck, a vehicle that looks like it was designed by a polygon-obsessed lunatic. You start up the infotainment system, anticipating the dulcet tones of your meticulously curated playlist. But instead of music, you're greeted by what can only be described as the spinning wheel of despair. The loading screen appears, a mocking, spinning circle that seems to say, "Patience, young Padawan."

This isn't your run-of-the-mill loading screen. Oh no. This is a loading screen with aspirations to be more. It's the Sisyphus of loading screens, eternally rolling its digital boulder up a never-ending hill. Minutes tick by. The screen remains unchanged, a monument to your dwindling patience.

As you sit there, staring into the abyss of the loading screen, you start to question your life choices. What led you to this moment? Why did you eschew the simple pleasure of a car with a straightforward, functional stereo for this torment? The loading screen becomes a portal for existential contemplation, a black hole sucking in your sanity.

But let's not dwell on the philosophical implications. Let's talk practicalities. You try the usual tricks – turning the system off and on, disconnecting and reconnecting your phone, uttering a string of increasingly creative expletives. Nothing works. The loading screen remains, steadfast in its mission to drive you to the brink of madness.

You consider going old school, humming to yourself or resorting to the archaic practice of conversation. But the allure of your playlist, with all its unplayed songs, is too strong. The loading screen has become your nemesis, a digital dragon guarding the treasure of your music.

As time stretches on, the cabin of the Cybertruck begins to feel less like the interior of a cutting-edge vehicle and more like a purgatory for the technologically damned. The promise of a playlist full of your favorite tunes hangs in the air, just out of reach, an unattainable nirvana.

Finally, in a moment that feels more miraculous than the parting of the Red Sea, the loading screen disappears. Music fills the cabin. The relief is palpable, a wave of joy washing over you. But the victory is tinged with paranoia. You eye the infotainment system warily, knowing that at any moment, it could revert to its default state of loading purgatory.

In conclusion, the eternal loading screen of the Tesla Cybertruck's infotainment system is more than a mere technological hiccup. It's a trial, a test of your mental fortitude and emotional resilience. It teaches you patience, endurance, and the value of silence – because sometimes, silence is all you have when your playlist is perpetually three seconds away from playing. In the Cybertruck, patience isn't just a virtue; it's a necessity. Welcome to the future, where your music is always just a little bit out of reach.

## Navigating the Abyss of the User-Unfriendly Interface

In the world of the Tesla Cybertruck, where every design seems to be a challenge against common sense, the infotainment system's user interface stands out as a beacon of user-unfriendliness. It's as if it was designed by someone who believes that ease of use is for the weak. This isn't just a user interface; it's an obstacle course designed by a sadist.

As you sit in the driver's seat, surrounded by more angles than a geometry textbook, you're faced with the daunting task of simply trying to change the radio station or, heaven forbid, adjust the climate settings. The interface greets you with a level of complexity that would baffle even the most seasoned cryptologist. Buttons, sliders, and menus are scattered across the screen with no apparent logic or order, like they were arranged by a tornado in an electronics store.

Attempting to navigate this labyrinthine mess is akin to playing a game of digital Twister. You want to adjust the temperature? That requires a swipe here, a tap there, and perhaps a small sacrifice to the tech gods. Trying to pair your phone via Bluetooth? You'd have better luck teaching Morse code to a pigeon. The options and settings are buried in submenus that are nested within submenus, like a Russian doll of frustration.

Then there's the responsiveness of the touchscreen, or rather, the lack of it. Each touch is met with a delay, a moment of suspense where you wait to see if your command has been registered or lost in the digital void. It's like playing a slot machine – you pull the lever and hope for the best, but more often than not, you're left with the spinning wheel of despair.

The design of the interface itself is a study in confusion. The color scheme seems to have been chosen by someone who thought 'drab' was the new 'black'. The fonts are either too small to be legible or so large they scream at you from the screen. It's a visual cacophony, an assault on the senses that leaves you longing for the simplicity of a rotary phone.

And let's not forget the voice command system, a feature that promises to alleviate some of the pain of using the touchscreen. However, this system seems to have been programmed to understand only a dialect of English spoken in a parallel universe. You ask it to play The Beatles, and it responds with a weather report for a city you've never heard of. It's like having a conversation with a well-meaning but hopelessly confused parrot.

Navigating through your media library is no less challenging. The system organizes your music in a manner that defies logic. Albums are jumbled, artists are mixed, and playlists seem to rearrange themselves every full moon. Trying to find the song you want to play is like trying to find a needle in a haystack, if the needle keeps moving and the haystack is in a labyrinth.

And just when you think you've mastered this digital Rubik's cube, Tesla releases an update, rearranging everything you just got accustomed to. It's like finally learning the rules to a game only to have someone change them mid-play.

In conclusion, navigating the user-unfriendly interface of the Tesla Cybertruck is a journey into madness. It's a test of patience, a trial of your problem-solving skills, and a challenge to your sanity. But perhaps that's the point. In the future, as envisioned by Tesla, ease of use is an outdated concept. Instead, we are expected to adapt, to evolve, to become one with the machine – or at least spend so much time trying to figure it out that we forget we're just trying to change the radio station. Welcome to the abyss of the user-unfriendly interface, where simplicity goes to die and confusion reigns supreme.

# A Maze of Menus: Delve into the labyrinthine interface

In the Cybertruck, Tesla has boldly reimagined the car infotainment system not as a tool for convenience, but as a labyrinthine puzzle that would have Theseus throwing up his hands in defeat. Navigating through its menus is an experience that makes less sense than following a drunkard's directions through the streets of Venice.

First, let's talk about the home screen – a place that should be a safe haven, a beacon of familiarity. In the Cybertruck, however, it's more like walking into a party where you don't know anyone, and everyone stops talking as you enter. The screen is a hodgepodge of icons, widgets, and information, arranged with the care and attention you'd expect from a toddler doing a jigsaw puzzle.

Attempting to change something simple, like the radio station, becomes an odyssey through this digital quagmire. You tap what you think is the radio icon, only to be taken to a GPS map of the Bolivian rainforest. You go back, try another icon, and suddenly you're adjusting the settings for the heated seats, which is about as useful as a chocolate teapot when all you want is some music.

The navigation menu deserves a special mention. It appears to have been inspired by the labyrinth of Crete, minus the Minotaur, although you wouldn't be surprised to find one lurking behind the submenu for Points of Interest. Entering a destination requires a degree in hieroglyphics, patience of a saint, and perhaps a small blood sacrifice.

Then there's the settings menu. Oh, the settings menu. It's a place of mystery and madness, where options are buried beneath options, and you find yourself longing for the days of simple knobs and buttons. Adjusting something like the cabin lighting involves a descent into a sub-menu Dante would have baulked at, where you're faced with sliders and adjustments that seem to operate in a different dimension of reality.

Let's not forget the climate control menu, a system that seems to operate on the principle that you should never be too comfortable. Adjusting the temperature involves a complex series of swipes and taps, and just when you think you've got it right, the system decides to reset itself, plunging you back into either an Arctic chill or Saharan heatwave.

The media player is another labyrinth within this maze. It's like trying to find your favourite song in a record store where the albums are arranged not alphabetically, not by genre, but by the phase of the moon when they were released. You scroll and scroll, passing albums and artists you've never heard of, and when you finally find what you're looking for, you tap it, and the system freezes, contemplating its existence.

And then there's the voice control – Tesla's answer to the maze of menus. Except, the voice control is like an overly literal genie, misinterpreting your commands with a mix of malice and incompetence. You ask it to play The Beatles, and it sets a course for Liverpool. It's less a convenience and more a cruel joke, a digital jester laughing at your attempts to navigate this digital Gordian knot.

In conclusion, the infotainment system in the Tesla Cybertruck is less a tool of convenience and more an exercise in digital masochism. It's a maze of menus, a labyrinthine interface that makes less sense than a drunkard's directions. Navigating through it requires patience, a strong will, and perhaps a sense of defeatism. But, like any good puzzle, once you've cracked it, once you've unravelled the mysteries of this digital enigma, you're left with a sense of accomplishment, of victory. That is until Tesla releases an update, and you're back to square one. Welcome to the maze; may you find your way out before madness takes hold.

## The Mystery of the Missing Features

In the world of the Tesla Cybertruck, the owner's manual is less an accurate guide and more a work of optimistic fiction. It promises a plethora of features, each more exciting than the last, creating a sense of wonder and anticipation. However, actually finding these features in the car is like participating in a treasure hunt where the treasure doesn't exist, and the map is a piece of abstract art.

Let's start with the so-called "advanced voice control" system. This manual sings songs of a future where your every spoken command is met with swift, accurate action from the Cybertruck. In reality, the voice control system has the comprehension skills of a profoundly confused parrot. You ask it to navigate to the nearest petrol station (a slip of the tongue, obviously, as it's electric), and it serenades you with a weather report for Timbuktu. It's less a convenience and more an exercise in futility and accidental comedy.

Then there's the elusive "self-cleaning" feature. Oh, how This manual boasts of a vehicle that maintains its own pristine condition. One would expect to return to a spotless car every time, but the reality is a vehicle that seems to attract dirt and grime like a magnet. Perhaps the self-cleaning feature is so advanced it's invisible, or more likely, it's just a figment of Tesla's aspirational imagination.

This manual also tantalizingly mentions a "fully-integrated espresso machine". This conjures images of sipping a freshly brewed coffee while admiring the dystopian, angular beauty of the Cybertruck's interior. However, the only thing brewing is disappointment, as the closest thing to an espresso machine in the Cybertruck is the cupholder. Maybe it's a feature only available in the parallel universe where the Cybertruck is a practical vehicle.

Let's not forget the "holographic display" – a feature that promises to project a 3D map of your surroundings, transforming your driving experience into something akin to piloting a spaceship. The sad reality is that the only thing holographic about the Cybertruck's display is its ability to project an illusion of Tesla's grand ambitions. The display is as flat as the humor in a bad sitcom, and the only thing it projects is conventional information in a decidedly conventional manner.

And then there's the "ambient sound generator", which, according to This manual, can mimic the sounds of various environments to enhance your driving experience. Want to feel like you're driving through a rainforest, complete with the sounds of exotic birds and trickling streams? Well, keep wanting, because the only ambient sound you'll hear is the low hum of disappointment and the occasional beep of an unclear, unhelpful sensor.

In conclusion, hunting for the features promised in the Cybertruck's manual is a journey filled with hope, anticipation, and ultimately, a sense of betrayal. It's like being promised a gourmet meal and getting a packet of crisps. The features, so boldly and vividly described in This manual, are as real as the Loch Ness Monster – often talked about, never seen. This manual, it seems, is an exercise in creative writing, a testament to Tesla's ambition, or perhaps their overconfidence. So, as you sit in your Cybertruck, surrounded by missing features and broken promises, remember – it's not just a car; it's an adventure into the world of what-could-have-been. Welcome to the mystery of the missing features. May your hunt be ever hopeful, and your disappointments, at least, be entertaining.

# 6. "Routine" Maintenance

In the fantastical realm of the Tesla Cybertruck, the term "routine maintenance" is as misleading as calling a tiger a "slightly upset kitty." It's a world where routine tasks become epic undertakings, requiring the patience of a saint and the mechanical understanding of a NASA engineer.

Let's start with something simple, like checking the oil. Oh wait, you can't. There isn't any. It's an electric vehicle. That's one less thing to worry about, or so you would think. But don't be fooled. The Cybertruck replaces the mundane oil check with a plethora of other checks that are about as routine as solving a Rubik's Cube blindfolded.

Checking the tire pressure should be straightforward, but in the Cybertruck, it's akin to deciphering ancient hieroglyphs. The pressure reading on the digital display fluctuates more than the stock market, leaving you wondering if you're about to embark on a journey with the correct tire pressure or if you'll end up as a roadside attraction.

And then there's the wiper fluid. In a normal vehicle, you pop the bonnet, find the reservoir, and fill it up. But in the Cybertruck, the process is shrouded in mystery. The reservoir is hidden beneath panels that require a special tool to remove – a tool, mind you, that's not included. It's like buying a jigsaw puzzle only to find out half the pieces are sold separately.

Replacing a headlight bulb, something that should be as easy as changing a light bulb at home, becomes a quest worthy of a Tolkien novel. This manual vaguely suggests it's possible, but actually doing it requires removing half the front end of the vehicle. You half expect to find a secret compartment with a dragon guarding it.

Updating the vehicle's software, which Tesla insists is routine maintenance, is a gamble. You initiate the update, and the vehicle becomes as responsive as a hibernating bear. The screen assures you it's only a few minutes. Hours later, you're left wondering if the Cybertruck has decided to retire from the world and live out its days as a very expensive driveway ornament.

Even cleaning the vehicle is a far cry from routine. The angular design of the Cybertruck, while visually striking, collects dirt and grime in its crevices like a street corner collects flyers. Washing it requires the precision of a surgeon and the dedication of a monk. You'll find yourself armed with a toothbrush, tackling each nook and cranny while questioning your life choices.

Checking the brakes is another "routine" task. This manual breezily suggests inspecting the brake pads and discs. But once you actually attempt this, you realize that the Cybertruck's brakes are as accessible as a hermit living atop a mountain. It's less a maintenance task and more an endurance test, challenging both your physical and mental fortitude.

In conclusion, routine maintenance on the Tesla Cybertruck is a misnomer. It's a series of trials, a collection of tasks designed to test your resolve, your ingenuity, and your will to persevere. It's not just maintenance; it's an adventure, a journey into the heart of the machine where you emerge not just as a vehicle owner, but as a battle-hardened warrior of automotive upkeep. So, when you roll up your sleeves to perform "routine" maintenance on your Cybertruck, remember – you're not just keeping your vehicle in good condition; you're embarking on an epic quest, worthy of song and story.

## Realigning the Wheels: A Guide to Guesswork

In the bizarre world of the Tesla Cybertruck, realigning the wheels is not so much a maintenance task as it is a dark art, akin to divination or reading tea leaves. In any sensible vehicle, this would be a straightforward process, perhaps involving a trained professional and some finely tuned equipment. In the Cybertruck, however, it's more of a wild guess, a foray into the unknown, where the tools of the trade are optimism and blind faith.

Let's start with the supposed 'simple' task of determining if your wheels actually need realigning. In a regular car, there might be tell-tale signs like uneven tire wear or the vehicle pulling to one side. In the Cybertruck, however, the signs are as clear as mud. The vehicle might start resembling a drunken sailor on leave, veering unpredictably. Or it might drive straight as an arrow, but wear down its tires faster than a politician backtracks on promises.

Once you've decided that, yes, your Cybertruck's wheels are about as aligned as the British government, you embark on the Herculean task of realignment. The owner's manual, a tome of optimism and vagueness, suggests this can be done with a few simple tools. What it fails to mention is that these 'simple' tools are akin to seeking the Holy Grail. You're better off fashioning your own out of whatever's lying around in your garage – perhaps that old set of wrenches and a prayer.

Now, assuming you've managed to source your mythical tools, the real fun begins. This manual provides instructions, but they're about as clear as a foggy day in London. Phrases like 'adjust the toe by turning the rod' and 'ensure the camber angle is correct' dance before your eyes, taunting your mechanical sensibilities. You feel like you're trying to crack the Enigma code.

Proceeding with the adjustment is a practice in guesswork and hope. You twist, turn, and tweak, guided only by your intuition and the vague instructions that might as well be written in ancient Sumerian. Every adjustment is a gamble – will this make the driving experience smoother, or will it result in a journey as shaky as a ride on a wooden roller coaster?

And let's talk about the alignment itself. In a sensible world, this would involve precise measurements and exact angles. In the Cybertruck, it's more a case of 'that looks about right'. You're aiming for a perfect 90 degrees, but there's a good chance you'll end up with something closer to a Salvador Dali painting – intriguing but fundamentally wrong.

Once you think you've finished, the moment of truth arrives. It's time to test drive. You take the Cybertruck out, heart pounding, palms sweaty, half-expecting the wheels to wobble like a clown car. The vehicle moves forward, and it's holding steady – a miracle! Or so it seems, until you hit the first corner and realize you've overcorrected, and now the Cybertruck handles like a supermarket trolley with a wonky wheel.

In conclusion, realigning the wheels of a Cybertruck is an exercise in patience, guesswork, and sheer bloody-mindedness. It's a process that tests your resolve, your mechanical skills, and your willingness to hurl expletives at an inanimate object. But, in a way, that's the beauty of the Cybertruck. It's not just a vehicle; it's a puzzle, a challenge, a riddle on four wheels. So, when you decide to embark on the quixotic quest of wheel realignment, remember – in the world of the Cybertruck, precision is just another word for 'good guess'. And at the end of the day, isn't that what adventure is all about?

## The Alignment Lottery

In the ever-baffling world of the Tesla Cybertruck, wheel alignment is akin to playing the lottery – except with worse odds and the prize is a straight drive. In any normal vehicle, wheel alignment is a precise operation, carried out by skilled technicians with sophisticated equipment. In the Cybertruck, it's more like a game of pin the tail on the donkey, played by blindfolded people who've never seen a donkey.

Let's set the scene. You're in your Cybertruck, a vehicle that looks like it was designed using a ruler and a protractor by someone who flunked art class. You've noticed that driving it straight is as challenging as explaining quantum physics to a toddler. The vehicle veers left and right with the unpredictability of a politician's promises. So, you decide it's time to check the wheel alignment – a decision you'll soon regret.

This manual, a hefty tome filled with optimism, suggests that aligning the wheels is a straightforward task. This is the first sign that you're entering a world of delusion. The Cybertruck, with its angular, brutish design, hides its alignment screws in locations so obscure, you're more likely to find Narnia than the adjustment points.

Armed with a toolkit that looks like it belongs to a space station, you set about trying to align the wheels. The process involves turning various screws and bolts, each adjustment bringing about as much predictability as throwing darts blindfolded. You turn a screw, hoping for improvement, only to find that the wheel now points towards the Andromeda Galaxy.

As you wrestle with levers and wrenches, the realization dawns that wheel alignment in the Cybertruck is less science and more witchcraft. You find yourself muttering incantations under your breath, hoping for divine intervention. Perhaps a quick prayer to the deity of mechanical engineering might not go amiss.

Then comes the moment of truth — the test drive. You set off, heart pounding with a mix of hope and fear. Initially, it seems you've triumphed — the Cybertruck is moving in a reasonably straight line. But as you gain speed, the harsh reality sets in. The vehicle starts to drift, veering with the grace of a drunken elephant. It's not alignment; it's anarchy.

Returning to the garage, you realize that wheel alignment in the Cybertruck is not a task; it's a saga. It's a journey through frustration, confusion, and despair. It's a battle between man and machine where the machine seems to be winning. Each turn of the wrench, each adjustment, is a roll of the dice — a spin in the great alignment lottery.

In conclusion, aligning the wheels of a Tesla Cybertruck is an adventure, a foray into the unknown. It's a task that tests your skill, your patience, and your sanity. It's a process that makes you long for the days of horse-drawn carriages, where the only alignment needed was a good carrot for the horse. But fear not, brave Cybertruck owner, for this is all part of the experience. Embrace the challenge, embrace the madness, and remember – in the alignment lottery, every spin of the wheel is a new opportunity for chaos. And isn't that, after all, part of the fun?

## When DIY Stands for Destroy It Yourself

In the brave new world of the Tesla Cybertruck, where every panel is sharper than Occam's razor, attempting DIY repairs is like trying to defuse a bomb with a hammer. It's not so much 'Do It Yourself' as it is 'Destroy It Yourself', a path fraught with danger, despair, and the very real possibility of turning a futuristic vehicle into a very expensive heap of scrap.

Imagine, for a moment, you decide to tackle a simple repair. Perhaps a headlight has gone out, and you think, "How hard can it be?" These famous last words are akin to saying, "What's the worst that could happen?" before opening Pandora's box. In a regular car, changing a headlight is as easy as changing a bulb in your living room. In the Cybertruck, it's more like performing keyhole surgery... blindfolded... in a hurricane.

You begin by consulting This manual, which is about as helpful as a chocolate teapot. The instructions, written in what can only be described as 'techno-babble', assume you have at your disposal the entire toolset of a SpaceX engineer and the knowledge to match. Undeterred, you set to work, armed with nothing but a screwdriver and unbridled optimism.

As you remove the first panel, a sense of unease creeps in. It comes off a little too easily, like a scab that wasn't ready to be picked. You peer into the abyss you've opened, a tangle of wires and components that look less like part of a car and more like an abstract art installation.

Progress is slow and treacherous. With each component you remove, the task seems to grow more complex. It's like playing a game of mechanical Jenga. Every piece removed increases the likelihood of the entire thing collapsing into a heap of broken dreams and warranty voids.

Halfway through, the Cybertruck looks less like a vehicle and more like a disassembled robot from a low-budget sci-fi movie. Parts and tools are strewn around like the aftermath of a small, localized tornado. You stand back and realize you have no idea how to put it back together. It's a jigsaw puzzle where all the pieces are grey and none of them fit.

But let's say, through sheer luck and desperation, you manage to replace the headlight and reassemble the Cybertruck. You turn on the engine, and for a moment, everything seems fine. That is until you turn on the headlights, and instead of a beam of light, the horn blares. Somehow, in your quest, you've rewired the horn to the headlight switch. It's a feature, not a bug, you tell yourself.

In conclusion, attempting DIY repairs on the Tesla Cybertruck is a cautionary tale, a tale that serves as a reminder that sometimes, leaving things to the professionals is not cowardice; it's common sense. It's a reminder that while ambition is admirable, ambition without knowledge is like a car without wheels – it's not going to get you very far. So, the next time you feel the urge to fix something on your Cybertruck, remember: DIY in this case might just stand for Destroy It Yourself. And maybe, just maybe, that's an adventure best left unexplored.

## Battery Maintenance: A Shocking Surprise

In the electric wonderland that is the Tesla Cybertruck, the concept of battery maintenance is as shockingly surprising as finding out the Queen is a secret heavy metal fan. One would think that in a vehicle more futuristic than a sci-fi movie set, maintaining the battery would be as easy as charging your mobile phone. Oh, how delightfully wrong you'd be.

First off, let's talk about accessing the battery. In a normal car, lifting the bonnet gives you access to the engine and, more importantly, the battery. In the Cybertruck, accessing the battery is like trying to break into Fort Knox with a spoon. This manual, written in what can only be presumed as an attempt at humor, suggests various panels and covers need to be removed. This requires a tool collection that would make a Formula One pit crew blush and a degree in mechanical engineering.

Once you've managed to locate and access the battery, nestled deep within the bowels of the Cybertruck like a bear hibernating in a cave, you're faced with a sight that would make a grown mechanic weep. Wires, cables, and components are arranged in what appears to be a deliberate attempt to confuse and intimidate. It's less a battery compartment and more a three-dimensional puzzle designed by a sadist.

Now, This manual breezily suggests checking connections and ensuring everything is secure. This is about as easy as performing a heart transplant with a pair of chopsticks. Each connection is a riddle, wrapped in a mystery, inside an enigma. You half expect to find a little note from Elon Musk himself, laughing at your naivety.

Charging the battery, something that should be as simple as plugging in a charger, becomes a game of Russian roulette. The charging port, hidden behind a panel so flush with the body it would make a stealth bomber jealous, is revealed by performing a sequence of swipes on the touchscreen. Get it wrong, and you might accidentally initiate the self-destruct sequence. Okay, maybe not, but it certainly feels that way.

The Cybertruck's battery management system is supposed to be state-of-the-art, a marvel of modern technology. In practice, it's about as predictable as a game show. Sometimes it charges quickly, other times it seems to take as long as waiting for the next season of your favorite show. The range indicator fluctuates with the whimsy of a mood ring, leaving you perpetually uncertain whether you have enough charge to get to the corner shop, let alone on a long journey.

This manual also warns of the dangers of overcharging, undercharging, and generally upsetting the battery in any way. It's like the battery is an overly sensitive artiste that must be coddled and reassured constantly. Treat it wrong, and it'll sulk, leaving you stranded with as much power as a wind-up toy.

In conclusion, battery maintenance in the Tesla Cybertruck is a shocking surprise, a journey into the unknown that tests your skills, your patience, and your sanity. It's an adventure, a challenge, a chapter in the ongoing saga of electric vehicle ownership. So when you roll up your sleeves and prepare to dive into the world of battery maintenance, remember – you're not just looking after a power source. You're nurturing the heart of a beast, a temperamental, complicated, and utterly baffling heart that keeps the Cybertruck alive. Good luck, you'll need it.

## Is your battery charging, or is it just pretending?

In the Tesla Cybertruck, the act of charging the battery is akin to participating in a historical reenactment of the Charge of the Light Brigade, complete with the same levels of confusion, misdirection, and a vague sense of impending doom. It's a process shrouded in mystery, optimism, and a nagging feeling that the vehicle might just be having a laugh at your expense.

Let's begin with the simple act of plugging in the charger. In any sensible vehicle, this would be as straightforward as plugging in a toaster. In the Cybertruck, it's more like trying to solve a Rubik's Cube that's fighting back. The charging port, hidden behind a panel that's smoother than a politician's excuse, reveals itself after a series of swipes on the touchscreen that feel more like a secret handshake than a functional gesture.

Once connected, you'd expect the simple confirmation that your vehicle is charging. But no, the Cybertruck decides to turn this into a game. The display shows a series of vague icons, animations, and numbers that seem to change with no discernible pattern or reason. It's like watching a stock market ticker; you know it's important, but you're not entirely sure what it's telling you.

The charging rate is a mystery in itself. Sometimes, the Cybertruck charges with the enthusiasm of a child on Christmas morning. Other times, it's more like it's decided to hibernate for the winter. There's no rhyme or reason. This manual suggests various optimal conditions for charging, but they read more like a recipe for a witch's brew than any practical advice.

Now, let's talk about the range. The Cybertruck, when fully charged, promises a range that seems impressive. However, this number is about as reliable as a weather forecast in the British Isles. Drive normally, and the miles start ticking down faster than the countdown on New Year's Eve. Turn on the air conditioning or dare to accelerate a bit briskly, and you can almost see the battery gauge wince and recede.

Monitoring the charging process becomes a hobby in itself. The estimated time to full charge displayed on the screen is a work of optimistic fiction. It changes so often you wonder if it's being calculated by a room full of monkeys with calculators. You start adjusting your plans, not based on actual time, but on Tesla time – a mysterious dimension where the normal rules of hours and minutes don't apply.

The real fun begins when you unplug the charger. The Cybertruck, now supposedly full of electric vigour, displays its range. But as you start driving, you realize that this number is more of a guideline, an aspirational figure, rather than an actual representation of reality. The range anxiety kicks in, turning every journey into a nail-biting adventure where the question "Will I make it?" adds an unwanted layer of excitement to your day.

In conclusion, charging the Tesla Cybertruck is not just a matter of plugging in a cable. It's an exercise in trust, patience, and often, blind optimism. It's a battle of wits between you and the vehicle, a technological tango where the steps are never quite clear. So, as you plug in your Cybertruck and watch the enigmatic dance of numbers and icons on the display, remember – you're not just charging a battery. You're participating in the Charge of the Light Brigade, a brave foray into the unknown, where the line between charging and pretending is as blurred as the boundaries of common sense in this electric steed.

## Jump-Start Jamboree

In the curious world of the Tesla Cybertruck, the simple task of jump-starting your vehicle is transformed into a spectacle I like to call the Jump-Start Jamboree. It's less about reviving a flat battery and more about discovering if your Cybertruck is going to graciously accept the help or throw a tantrum that would put a tired toddler to shame.

First, let's set the scene. You've ignored all the warnings, pushed the limits of the battery, and now your Cybertruck sits as lifeless as the career of a reality TV star. No problem, you think. A quick jump-start should do the trick. But in the world of the Cybertruck, nothing is quick, and nothing is simple.

You haul out the jumper cables, which are less cables and more like two anacondas that have seen better days. Connecting them requires a degree in electrical engineering and the flexibility of a contortionist. This manual vaguely suggests connecting red to red and black to black, but it fails to mention that doing so in the Cybertruck is akin to defusing a bomb – one wrong move, and it's lights out, possibly permanently.

Once you've connected the cables, braving sparks that resemble a Fourth of July fireworks display, it's time to start the donor vehicle. This is where the fun really begins. As the donor car revs, the Cybertruck reacts like it's just been offered a sip of the finest champagne – it either accepts with grace or spits it back in your face.

If you're lucky, the Cybertruck's display flickers to life, a symphony of lights and beeps that suggests you might just have pulled it off. But then, the real performance begins. The lights on the dashboard start dancing, flickering, and flashing in a pattern that suggests the Cybertruck is either very grateful or about to launch into space.

Attempting to start the Cybertruck now is a gamble. Will it roar to life, or will it simply produce more lights and a series of concerning clicks? You turn the key (or, in this case, push the button), and the vehicle shudders, groans, and then... nothing. It's like waking a hibernating bear – you might get a response, or you might get a paw in the face.

But let's say the stars align, and the Cybertruck decides it's not its time to ascend to the great charging station in the sky. The engine (or whatever sorcery powers this beast) hums to life, and you breathe a sigh of relief. Your Cybertruck, against all odds, has been resurrected.

In conclusion, the Jump-Start Jamboree with a Tesla Cybertruck is less a roadside repair and more an exercise in electrical brinkmanship. It's a test of your mechanical skill, your patience, and your willingness to dance with danger. Jump-starting a regular car is as easy as making a cup of tea. Jump-starting a Cybertruck is like making a cup of tea in a hurricane – while blindfolded – and the teapot is on fire. So, the next time you find yourself in need of a jump-start in your Cybertruck, remember: you're not just kick-starting a car; you're participating in a ritual, a ceremony, a jamboree that's equal parts terror and triumph. And isn't that, after all, part of the joy of owning such a beast?

## Software "Upgrades": Crossing Your Fingers and Toes

In the enigmatic world of the Tesla Cybertruck, software "upgrades" are akin to participating in a high-stakes game of digital roulette. Each update is not so much an enhancement as it is a leap into the unknown, where you cross your fingers, toes, and possibly your eyes in the hope that your beloved truck doesn't decide to reinvent itself as a two-tonne paperweight.

Embarking on a software upgrade in the Cybertruck is like opening a mystery box from an eccentric billionaire. You're hopeful, but you're also acutely aware that you might unleash something that will have you hurtling down the motorway, screaming for mercy. The 'Install Now' button glows ominously on the screen, like the red button in a nuclear bunker – you know you shouldn't press it, but the temptation is overwhelming.

Once you've initiated the upgrade, the truck enters a state of electronic hibernation. The screens go dark, the usual hum of electric life ceases, and you're left staring at a hulking mass of uncertainty. It's during these moments that you reflect on your life choices, pondering why you didn't just buy something simple and understandable, like a horse.

As the minutes tick by, you begin to wonder if the Cybertruck has decided to upload its consciousness to the cloud, leaving its earthly shell behind. The progress bar on the screen moves with the speed and enthusiasm of a snail crossing a particularly sticky patch of road. It's a masterclass in patience, a test of your ability to withstand the creeping tide of technological anxiety.

Eventually, the screens flicker back to life, heralding the completion of the upgrade. But the relief is short-lived. Now comes the moment of truth — discovering what the update has actually done. In an ideal world, it would bring new features, fix old bugs, and generally enhance your driving experience. In the Cybertruck, it's more like spinning the wheel on a game show — you might win a holiday, or you might get a pie in the face.

You start to explore the new software, and it quickly becomes apparent that Tesla's idea of an upgrade is to move everything around, change how basic functions work, and introduce a whole new set of mysterious icons that resemble ancient hieroglyphs. It's like coming home to find someone has rearranged your furniture, painted your walls a different color, and replaced your cat with a slightly different cat.

Some features have vanished entirely, spirited away in the dead of night like a garden gnome in a neighborhood prank. Others have appeared, unannounced and unexplained, like distant relatives at a wedding. This manual, already as clear as a foggy day in London, is now about as useful as a chocolate teapot, offering no clues as to what these new changes entail.

Navigating this brave new world is a challenge. You press what you think is the button for the climate control, only to find you've opened the sunroof. You try to change the radio station and somehow engage the autopilot. It's less a car and more a carnival ride, where the only thing you can do is hold on and hope you don't crash.

In conclusion, software upgrades in the Tesla Cybertruck are not for the faint of heart. They are a digital adventure, a foray into the unknown, a test of your resolve against the whims of Tesla's software engineers. Each upgrade is a journey, a story, a saga of man versus machine, where the outcome is as unpredictable as the British weather. So, the next time your Cybertruck prompts you to install an upgrade, take a deep breath, cross your fingers and toes, and prepare for a ride into the uncharted territory of Tesla's latest whims. Welcome to the future; it's confusing, but it's never dull.

## Update or Downgrade

In the perplexing world of the Tesla Cybertruck, every software update is a spin of the Russian roulette wheel. It's an experience filled with suspense, much like watching a suspense thriller, but instead of sitting comfortably on your sofa, you're perched in a vehicle that cost you an arm and a leg. The anticipation isn't about whether the hero will save the day; it's about whether this update will transform your futuristic marvel into an oversized, very expensive driveway ornament.

When the notification for a software update pings, it brings with it a sense of foreboding. Updating your Cybertruck is like playing a high-stakes game of chance. Will it give you that promised increase in efficiency, or will it transform your UI into something resembling an abstract expressionist painting? The thrill of this gamble isn't for the faint-hearted. It's like expecting to watch a serene nature documentary and instead finding yourself in the middle of an action-packed, high-speed car chase.

You hit 'Update', and the process begins. The Cybertruck becomes as responsive as a sullen teenager. The screens go blank, and a small loading icon appears, spinning endlessly like a hypnotist's wheel. It's a moment of eerie silence, a digital deep breath before the plunge. You watch the loading bar with the same intensity as a football fan during a penalty shootout.

Then, as the progress bar reaches completion, the car reboots. The screens light up, and you're greeted with... well, it's hard to say, really. The layout of the infotainment system has changed, resembling the aftermath of a toddler's temper tantrum. Icons and menus are scattered haphazardly, as if the design team decided to throw everything into the air and see where it landed. You try to find the settings menu, but it's now playing hide and seek, apparently having developed a fondness for obscurity.

Some 'improvements' are immediately noticeable. The voice command system, previously only mildly disobedient, now seems to have taken an advanced course in misunderstanding human speech. You ask it to play The Beatles, and it responds with a cheerful weather report from a city halfway across the globe. It's less an AI assistant and more a random response generator.

Then there's the navigation system. Before the update, it was a reliable, if somewhat dull, companion. Post-update, it's decided to embrace adventure. The shortest route between two points might still be a straight line, but your Cybertruck now prefers scenic detours, roundabout routes, and the occasional off-road escapade. It's not getting lost; it's an 'unexpected adventure feature', according to the release notes.

But it's not all doom and gloom. Sometimes, an update brings a genuinely useful feature – like an improved battery range or more efficient energy consumption. These are the moments that restore your faith, the software equivalent of finding a twenty-pound note in an old pair of jeans. It's a pleasant surprise, a ray of sunshine in the otherwise cloudy world of Cybertruck updates.

In conclusion, each software update for the Tesla Cybertruck is a leap into the unknown. It's a blend of excitement and apprehension, hope and fear. Will this update be a step forward into a brighter, more efficient future? Or will it be a step back into a realm of new bugs and electronic gremlins? You never really know, and perhaps that's part of the charm. It's not just a software update; it's an adventure, a story with each new chapter bringing unexpected twists and turns. So buckle up, hit 'Update', and enjoy the ride. After all, in the world of the Cybertruck, the only thing you can expect is the unexpected.

## The Eternal Reboot

In the realm of the Tesla Cybertruck, where technology is as advanced as a space shuttle and about as reliable as a chocolate teapot, the eternal reboot is a feature that transforms the simple act of starting your car into a suspense-filled drama. It's a process that allows you to ponder the mysteries of the universe, or at least the mysteries of why you thought owning a vehicle that's essentially a computer on wheels was a good idea.

Imagine this: You get into your Cybertruck, a vehicle that looks like it was designed by a geometry teacher with a vendetta against curves. You press the start button, expecting the usual hum of electric readiness. Instead, you're greeted with the spinning wheel of doom on the touchscreen, indicating that the Cybertruck has decided to take an impromptu digital nap.

Now, in any normal car, a reboot takes a few moments. In the Cybertruck, it's an opportunity to sit back and watch as your vehicle ponders the very essence of its existence. The screen flickers, the dashboard lights perform a disco dance, and you're left wondering whether you've accidentally initiated self-destruct.

This is the perfect time to enjoy a cup of tea. Why, you ask? Because the Cybertruck's reboot has all the urgency of a sloth crossing the road. You can boil the kettle, brew your tea, have a biscuit or two, and still return to find the vehicle in a state of electronic contemplation. It's less a car and more a Zen master, teaching you the virtues of patience and the art of stillness.

As you sip your tea, you might ponder the various systems and features in the Cybertruck, all currently in a state of suspended animation. The advanced navigation system, the infotainment console, the climate control – all taking a break from their digital labors. It's a reminder that in our fast-paced world, sometimes even machines need a rest. A rest that's about as convenient as a flat tire on a rainy day.

The eternal reboot is not just a waiting game; it's a gamble. Each time the system restarts, it's a roll of the dice. Will it return to its usual state of electric grace, or will it emerge with a new set of quirks and idiosyncrasies? Will the GPS now think you're in rural Mongolia? Will the voice control start speaking in tongues? The possibilities are endless and endlessly entertaining.

And just when you think it's done, when the screen finally displays the familiar interface, the Cybertruck may throw a curveball. Perhaps it decides to reboot again, just for good measure, plunging you back into the limbo of technological uncertainty. It's like being in a TV show where the plot twist is that there is no plot, just an endless cycle of reboots and tea.

In conclusion, the eternal reboot of the Tesla Cybertruck is more than a mere inconvenience; it's a journey, an experience, a lesson in the virtues of patience and the impermanence of technology. It teaches you to embrace the unexpected, to find joy in the little things – like the time it takes to enjoy a good cup of tea. So the next time your Cybertruck decides to ponder the mysteries of the digital universe, remember to relax, have a biscuit, and enjoy the ride – however stationary it might be. In the world of the Cybertruck, the journey doesn't always involve moving forward. Sometimes, it's about sitting still and savoring the moment, one reboot at a time.

# 7. Emergency Pretenses

In the fascinating world of the Tesla Cybertruck, the concept of an emergency takes on a whole new dimension. It's like being in a Bond movie, but you're not quite sure if you're the suave 007 or the villain whose car inevitably explodes. The Cybertruck, with its fortress-like appearance, gives you a sense of invincibility, but when it comes to emergencies, it's more like a deer in headlights – startled, confused, and not particularly helpful.

Let's begin with the basics – the emergency brake. In most vehicles, this is a simple, mechanical system. Pull a lever or press a pedal, and you stop. In the Cybertruck, however, the emergency brake is more akin to a suggestion to the car. You engage it and hope it's in the mood to listen. Sometimes it responds with the urgency of a cat being told to get off the sofa. Other times, it seems to enter a state of existential dread, pondering why it exists at all.

Then there's the matter of changing a flat tire. In this mammoth of a vehicle, you'd expect a flat tire to be an event, complete with warning lights, alarms, and maybe a comforting voice telling you everything will be okay. Instead, the Cybertruck offers a small, almost apologetic icon on the display. It's less of an alert and more of a shy whisper, easily missed in the cacophony of digital noise on the dashboard.

And should you decide to change the tire yourself, you're in for a treat. The Cybertruck's tires, massive and imposing, look like they belong on a lunar rover. Changing one is a Herculean task that would leave Hercules himself scratching his head. The jack provided is about as useful as a chocolate fireguard, and the process of removing the tire is akin to wrestling an angry gorilla.

In the event of a breakdown, the Cybertruck's computer system decides to be 'helpful'. It offers a plethora of diagnostics, graphs, and data, none of which makes any sense unless you have a PhD in Cybertruckology. The screen displays various error messages, none of which correlate to the actual problem. It's like asking a toddler why they're crying – you receive a lot of noise, but no useful information.

If you're unfortunate enough to have an accident, the Cybertruck transforms from vehicle to fortress. The doors lock, the windows seal, and you're ensconced in a high-tech cocoon. On paper, this sounds reassuring. In reality, it feels like you're in a panic room, with the added panic of trying to figure out how to get out. This manual, in its ever-helpful manner, suggests various procedures, but in the heat of the moment, it's about as useful as a screen door on a submarine.

Calling for assistance using the Cybertruck's interface is another adventure. The system, designed to be intuitive, takes you on a merry dance through menus and submenus. By the time you've finally managed to call for help, you've had enough time to learn basic car mechanics and fix the problem yourself.

And let's not forget the first aid kit. Hidden in the bowels of the truck, accessing it is a puzzle that would stump Indiana Jones. By the time you've retrieved it, you're left wondering whether the injury might just heal on its own.

Emergency pretenses in the Cybertruck, therefore, are a mix of high drama and low practicality. It's like being in a thriller where the hero forgets his lines. You're promised a vehicle that can handle anything, but at the first sign of trouble, it turns into a diva, demanding attention, offering confusion, and generally being about as helpful as a glass hammer. So, when you find yourself in an emergency in the Cybertruck, remember – you're not just in a sticky situation; you're in an episode of a sitcom, where the car is the comedian, and you're the straight man.

## How to Utilize the Decorative Parachute

In the otherworldly realm of the Tesla Cybertruck, where every feature is as practical as a chocolate teapot, we find the decorative parachute, a feature that is as useful as an ashtray on a motorbike. Now, you might wonder, why on earth does a terrestrial vehicle, which moves about as fast as a tortoise wading through treacle, need a parachute? Well, it doesn't, but in the land of Tesla, practicality is often thrown out of the window, which in this case, is bulletproof.

Firstly, let's address the elephant in the room – the parachute is decorative, which means it's as functional as a screen door on a submarine. You might as well be trying to utilize a painting of a fire extinguisher in an inferno. However, let's not let reality dampen our spirits. Let's explore how one might 'utilize' this glorified piece of fabric.

This manual, which has a tenuous relationship with reality at the best of times, suggests the parachute could be used for 'aesthetic aerodynamics'. Now, as anyone with a basic grasp of physics will tell you, this is about as sensible as using a feather to anchor a ship. The notion that a piece of fabric flapping off the back of your truck could somehow improve its aerodynamics is laughable. It's like claiming a top hat will make you run faster.

But let's indulge this fantasy for a moment. Say you decide to deploy this decorative parachute while cruising down the motorway. Aside from the bewildered looks from other drivers, who are probably wondering if a circus act has just escaped, you'll likely find the only thing it achieves is to act like an oversized airbrake. The Cybertruck, already not the nippiest vehicle on the road, will slow to a speed that makes glaciers seem positively hasty.

Perhaps the decorative parachute could be used as a picnic blanket. It's certainly large enough, and in a pinch, it might just serve as a decent ground cover. Of course, the material, designed to be lightweight and flutter in the breeze, will likely take off at the first hint of wind, turning your peaceful picnic into a scene reminiscent of Mary Poppins' less successful landings.

This manual also whimsically suggests that the parachute could double as a fashion statement. Now, unless you're attending a fancy dress party where the theme is 'rejected Bond gadgets', I struggle to see how this could be considered fashionable. Draping yourself in the parachute would be akin to wearing a tent. You'd be less haute couture and more 'haute camping'.

Let's not forget the potential use as a distress signal. Should you find yourself stranded, the bright fabric of the parachute could be used to attract attention. However, given its size, you're more likely to be mistaken for a strange, new species of flora by any passing aircraft. Rescue helicopters will circle, puzzled by the giant, colorful bloom on the landscape.

In conclusion, the decorative parachute of the Tesla Cybertruck is a solution in search of a problem. It's a feature that's about as useful as a submarine with a sunroof. But in the grand tradition of Tesla, it's quirky, it's different, and it gives you something to talk about at parties. So, if you ever find yourself bored and in possession of a decorative parachute, remember, it's not just a pointless piece of fabric; it's an adventure waiting to happen. Just don't expect it to be a particularly useful one.

**Understanding that the parachute is more of a concept than a reality.**

In the Tesla Cybertruck, you'll find an array of features that boggle the mind, and then there's the parachute. It's mentioned in This manual, a document that seems to be written by someone who believes the Cybertruck is capable of interstellar travel. This parachute, a feature more mysterious than the Bermuda Triangle, is touted as an essential safety device. But in reality, it's about as useful as a chocolate fireguard.

Now, you might scour the Cybertruck looking for this elusive parachute, perhaps imagining a James Bond-style escape mechanism, a ripcord you can yank to gracefully glide to safety from the clutches of mundane traffic. However, the stark reality is that the parachute is more of a concept, an idea, a whimsical fantasy of the Tesla engineers who probably chuckled into their kale smoothies as they added it to This manual.

The notion that a truck, a vehicle more suited to lugging garden waste than soaring through the skies, would need a parachute is laughable. It's like fitting a space rocket with a bicycle rack – technically possible, but utterly pointless. The Cybertruck, with its hefty frame and angular design, is about as aerodynamic as a brick. The idea that it could glide gently to the ground, cradled by a silken parachute, is the stuff of fairy tales.

Let's entertain the concept for a moment. Suppose you're cruising along, and for some unfathomable reason, you decide to deploy this mythical parachute. What then? Does the truck sprout wings? Does it gently lift off the ground, leaving other motorists gawking in awe? No. The reality is you'd likely trigger a series of confused error messages on the dashboard, and the most excitement you'd get is from the roadside assistance operator trying not to laugh as you explain the situation.

But let's not be too harsh. The parachute, while a work of fiction in the physical world, is a symbol, a metaphor for the audacity of the Cybertruck. It represents the boundless ambition, the relentless pursuit of innovation, the unyielding desire to push the boundaries of what a vehicle can be. It's a dream, a vision, a slice of a future where cars don't just drive – they fly.

This manual, in its infinite wisdom, fails to clarify that the parachute is not an actual, tangible thing. It's an idea, a piece of poetic license, a nod to the adventurous spirit that the Cybertruck embodies. It's Tesla's way of saying, "Yes, we put an electric motor in a pick-up truck, why wouldn't we add a parachute?"

In the end, the parachute that never was is a lesson in expectation versus reality. It teaches us that not everything you read in This manual should be taken at face value. It reminds us to dream, to imagine, to think outside the box, even if that thinking results in the idea of a parachute in a truck. So, the next time you glance at that manual and see the mention of a parachute, remember – it's not a lie, it's not a mistake, it's a testament to human imagination. And in the world of the Tesla Cybertruck, imagination is the only limit.

## False Hopes and Dreams

In the fantastical world of the Tesla Cybertruck, where design and practicality are as mismatched as socks on a teenager's bedroom floor, certain features give off the illusion of usefulness. These are the automotive equivalent of a chocolate teapot – lovely to look at, but utterly futile when it comes to making tea. In this hallowed tome of deceit and disillusionment, let's guide you through the mirage of features that are there for show, teasing you with false hopes and dreams.

Let's start with the Cybertruck's exterior. It looks like it was designed using a ruler by someone who was vehemently against the concept of curves. You might think this angular design serves a purpose – perhaps it deflects wind or repels dirt like a non-stick frying pan. No. In reality, these angles serve as convenient catchments for every speck of dirt, leaf, and unfortunate insect, turning the truck into a rolling exhibit of the local ecosystem.

Moving inside, the touchscreen is as large as a small country's GDP. It promises an interface so intuitive, you'd think it could read your thoughts. But let's be honest, it's more likely to misinterpret your gentle tap as an aggressive jab, sending you to a map of the Andes instead of adjusting the air conditioning. It's less a center of control, more a center of confusion, a smorgasbord of options and features presented in a way that would puzzle even the Enigma code breakers.

Then there's the voice control system, a feature that promises to transform your spoken word into vehicular action. In theory, it's like having Jeeves in your car, ready to attend to your every need. In practice, it's more like having a confused parrot, eagerly listening but ultimately squawking back something unintelligible. Ask it to play Mozart, and you might as well brace yourself for a blast of heavy metal.

Oh, and let's not forget the supposedly 'self-tightening' bolts, a feature that conjures up images of little robotic arms inside the vehicle, diligently keeping everything in check. The reality is these bolts are about as likely to self-tighten as your sneakers are to self-tie. It's a concept that defies basic mechanics, leaving you to wonder if the Cybertruck was assembled in Narnia.

The Cybertruck's purported amphibious capabilities are another fantasy. The vehicle looks as though it could tackle a jaunt across the English Channel. However, the truth is, it's about as seaworthy as a sieve. Any attempt to transform your Cybertruck into a makeshift boat will leave you not so much sailing into the sunset, as sinking into the murky depths of despair.

And then there's the bulletproof claim. Now, while it's undeniably cool to say your truck can stop a bullet, the reality is you're more likely to encounter a rogue shopping cart than a stray round. The bulletproof feature is akin to wearing a suit of armor to a chess match – impressive, but entirely unnecessary and frankly, cumbersome.

In essence, the Tesla Cybertruck is a cornucopia of false hopes and dreams, a collection of features that promise the world but deliver an atlas. It's a vehicle that, at first glance, seems capable of conquering Mars, but in reality, struggles with the concept of a muddy field. The lesson here is to temper your expectations. Approach the Cybertruck not as a marvel of modern technology, but as a vehicle sprinkled with a liberal dose of optimism and a pinch of fantasy. It's not just a truck; it's a rolling testament to human ambition, where the line between reality and aspiration is as blurred as the ownership of a shared fridge. Just remember, when you climb into the Cybertruck, you're not entering a realm of limitless possibility, but a world where the line between the practical and the theatrical is as thin as the vehicle's paint job.

# Deploying the Theoretical Safety Bubble

Deploying the Theoretical Safety Bubble in the Tesla Cybertruck is an exercise in imagination and wishful thinking, much like trying to catch a unicorn in your garden. This manual, a tome of optimism and fantasy, suggests that in the event of a collision, a 'safety bubble' will deploy, enveloping the vehicle in a protective embrace. This, however, is akin to expecting a bubble wrap to stop a charging rhinoceros.

First, let's consider the physics of this so-called safety bubble. In theory, it's a marvel of engineering, a feat of technology that would make even NASA engineers green with envy. In practice, it's about as effective as a chocolate teapot. The idea that an invisible force field will pop up, like a magic shield in a video game, to protect you and your precious Cybertruck from the harsh realities of physics is, frankly, ludicrous.

Now, should you find yourself in a situation where the Cybertruck deems it necessary to deploy this mythical safety bubble, This manual advises you to remain calm and trust in the technology. This is the equivalent of telling someone on a sinking ship to just stop and admire the view. As you sit there, waiting for the bubble to deploy, you'll likely have enough time to reflect on your life choices, write a memoir, or perhaps learn a new language.

This manual also thoughtfully includes a troubleshooting section for the safety bubble, which is about as useful as a handbook on dragon taming. It suggests checking the sensors and ensuring they are free from dirt and debris. Of course, by the time you've done this, it's entirely possible that the need for the safety bubble has passed, and you're now simply sitting in a very expensive, very stationary, and decidedly unbubbled Cybertruck.

Let's not forget the visual spectacle of the safety bubble. In the imaginative world of This manual, it's a shimmering, translucent shield, a glowing orb of protection. In reality, should it ever decide to make an appearance, it's more likely to resemble a confused fog, briefly misting up the windows before dissipating into the ether.

This manual also fails to mention the potential reactions of bystanders and other motorists to your deploying safety bubble. In the unlikely event that it does activate, it's sure to cause more confusion and alarm than reassurance. Passersby might wonder if the Cybertruck is about to launch into space or if it's part of a poorly executed magic trick.

In essence, the theoretical safety bubble of the Tesla Cybertruck is a wonderful piece of science fiction, a delightful daydream of what might be possible in a world where the laws of physics are mere suggestions. It's a testament to the ambition and imagination of the designers, a whimsical feature in a vehicle that's part truck, part spaceship, and part fantasy.

As an owner of this vehicular enigma, it's important to remember that the safety bubble, while a charming concept, is about as reliable as a sundial at night. It's a promise of safety wrapped in layers of theoretical physics and wishful thinking. So, should you ever find yourself in a situation where the Cybertruck suggests deploying the safety bubble, it's probably best to remember that you're in a car, not a sci-fi movie. Keep your wits about you, your hands on the wheel, and maybe, just for good measure, cross your fingers. After all, in the Cybertruck, reality and fantasy are just two sides of the same coin, and who knows which side will land up in a pinch?

## Discover the safety features that exist only in This manual.

In the fantastical realm of the Tesla Cybertruck, the concept of safety is treated with the same level of seriousness as a clown at a birthday party. This manual, a whimsical document that could double as a work of speculative fiction, promises a smorgasbord of safety features. These features, however, are as elusive as a pleasant conversation about politics at a family dinner.

Let's dive into the deep end of this pool of make-believe safety. First, there's the mention of an "auto-avoidance system," a feature that supposedly detects potential collisions and takes evasive action. In theory, it's like having a superhero at the wheel, ready to leap into action at the first sign of danger. In practice, it's more akin to having a nervous squirrel in charge. The system is as likely to swerve you into danger as it is to steer you away from it, making every journey a thrilling toss of the dice.

Then there's the "advanced fire suppression system," a safety feature that conjures images of a high-tech response to any sign of flame. You might expect a futuristic foam to deploy at the first whiff of smoke. However, the only thing this system seems to suppress is your confidence. In the event of a fire, the system is more likely to give you a stern warning on the dashboard – "Caution: It's getting a bit warm" – while you frantically search for the nearest exit.

This manual also boldly claims the presence of "impact-absorbing body panels." These panels, made from a material apparently sourced from the planet Krypton, are supposed to cushion the blow from any collision. The reality is that these panels are about as impact-absorbing as a sheet of aluminum foil. In a crash, they're more likely to crumple into an origami sculpture than provide any real protection.

Let's not forget the "emergency beacon system," a feature that supposedly alerts emergency services with your precise location in the event of an accident. It sounds reassuring until you realize that it's about as effective as sending a carrier pigeon. You're more likely to be found by a passerby with a keen sense of direction than by relying on this imaginary beacon of hope.

The "solar-powered survival mode" is another gem. According to This manual, should you find yourself stranded, the Cybertruck can sustain life support systems indefinitely. This is a feature seemingly borrowed from a science fiction novel. In reality, the only thing this mode will sustain is a growing sense of despair as you realize that the solar panels are about as useful as a handbrake on a canoe.

And then there's the pièce de résistance: the "underwater escape function." The Cybertruck, claims This manual, can be driven underwater, turning it into a makeshift submarine. This is a feature that would make James Bond envious, if only it were true. In reality, attempting to drive the Cybertruck underwater would result in a very expensive, very electric, and very wet brick.

In essence, the safety features of the Tesla Cybertruck, as described in This manual, exist in a realm of fantasy. They are a collection of promises as reliable as a weather forecast in Britain. It's a veritable cornucopia of innovative ideas, none of which bear any resemblance to reality. So, as you strap yourself into this marvel of automotive ambition, remember that the safety features are more a flight of fancy than a parachute of protection. Your best bet is to drive cautiously, avoid danger, and maybe keep a pillow on your lap – just in case.

## Is it a safety bubble or just a bubble of hot air?

In the realm of the Tesla Cybertruck, where reality often takes a back seat to Elon Musk's imagination, resides the mythical safety bubble. This feature, more elusive than a politician's promise, is touted in This manual as the ultimate safety feature. But let's be honest, the only bubble here is the one the designers were living in when they thought this up.

The safety bubble, according to the gospel of Tesla, is a force field that envelops the Cybertruck in the event of a collision, protecting it like a mother hen guards her chicks. In theory, it's brilliant – a cocoon of safety, shielding you from the harsh realities of physics. In practice, it's about as effective as a paper umbrella in a hurricane. The idea that a bubble – essentially air – can protect you in a crash is the sort of thing you'd come up with after a night of heavy drinking and watching sci-fi movies.

Now, let's say for the sake of argument, you find yourself in a situation where the safety bubble would be handy. You brace for impact, expecting this wondrous bubble to spring into action. What happens? Well, the dashboard lights up like a Christmas tree, the Cybertruck emits a series of concerning beeps, and you're left sitting there, waiting for a bubble that never comes. It's like waiting for Godot, if Godot was a safety feature in a high-tech truck.

This manual also suggests that the safety bubble can deploy in the event of a rollover. The Cybertruck, with its low center of gravity, is about as likely to roll over as a sumo wrestler, but let's not dwell on facts. Should the unthinkable happen, and you find yourself upside down, the safety bubble is supposed to gently right the vehicle. The reality is you're more likely to get a symphony of warning sounds and a stern message on the touchscreen advising you to, "Please avoid inverting the vehicle."

But let's not be too cynical. Perhaps the safety bubble isn't a physical entity. Maybe it's a metaphor, a symbol of the Cybertruck's advanced safety systems – a comforting thought that envelops you as you hurtle down the motorway, secure in the knowledge that you're driving a vehicle that was designed with safety in mind. Or maybe it's just a fanciful notion, a bubble of hot air, a marketing gimmick in a vehicle that's as much about show as it is about substance.

The safety bubble also gives rise to an important philosophical question: If a safety feature exists in This manual but never deploys in real life, does it actually exist? It's the automotive equivalent of the sound of one hand clapping – a conundrum that's sure to keep you entertained as you sit in traffic, surrounded by the cold, hard steel of the Cybertruck, with nary a bubble in sight.

So, the next time you read about the Cybertruck's safety bubble, remember that it's more a figment of imagination, a bubble of wishful thinking. It's a concept that's as grounded in reality as the Cybertruck is capable of flying. And in the grand tradition of Tesla, it's a reminder that innovation often starts with an idea, a dream, a bubble – even if that bubble is made of nothing but hot air.

## The Ejector Seat: For Exiting Conversations About Mileage

In the wildly imaginative world of the Tesla Cybertruck, among the list of features that sound like they've been plucked straight from a James Bond film, resides the ejector seat. This is particularly useful for those moments when you're trapped in a dreary conversation about the vehicle's mileage at a dinner party. The theory is simple: hit a button, and you're launched skywards, free from the shackles of small talk and fuel efficiency debates.

Now, let's be clear. The inclusion of an ejector seat in a civilian vehicle is about as sensible as installing a fireplace in a submarine. It's the kind of feature that makes you wonder whether the designers at Tesla were given free rein to live out their childhood fantasies without the hindrance of adult supervision.

Imagine the scenario: you're at a garden party, and someone corners you to talk about the mileage of your Cybertruck. Their eyes glint with the intense curiosity of a cat staring at a laser pointer, and you know, deep in your soul, that this conversation is going to be as long and painful as a dentist appointment. In a normal car, you'd be forced to endure this torture. But in the Cybertruck, salvation is just a button press away.

The mechanics of the ejector seat are shrouded in mystery, much like the recipe for Coca-Cola or the true contents of a hot dog. This manual cryptically suggests it employs 'state-of-the-art propulsion technology'. In reality, this means you're relying on the same technology that powers those chairs at the funfair which shoot you up a tower – terrifying and exhilarating in equal measure.

Pressing the eject button initiates a sequence that's more dramatic than a season finale of your favorite TV show. First, the roof of the Cybertruck opens like the jaws of a giant mechanical beast. Then, with a whoosh that would put a NASA rocket to shame, you're catapulted upwards, leaving behind a trail of bewildered onlookers and a half-finished conversation about mileage.

Of course, what goes up must come down. The Cybertruck's manual glosses over this part with the vague assurance of a 'parachute for safe descent'. One can't help but imagine a scenario where you're floating back down to Earth, gently drifting into the neighbor's barbecue, or worse, landing in their swimming pool.

It's important to note that using the ejector seat for escaping mundane conversations does come with its risks. Aside from the obvious danger of hurtling through the air at breakneck speeds, there's also the social fallout. It's a surefire way to make an exit, but you may not be invited back for the next soiree. And explaining to the paramedics how you ended up stuck in a tree while trying to avoid a chat about electric car mileage might be a tad awkward.

So, while the ejector seat in the Tesla Cybertruck may seem like the answer to all your social escape needs, it's perhaps best left as a last resort. After all, enduring a tedious conversation about mileage is preferable to explaining to your insurance company why you felt the need to eject yourself from a stationary vehicle in the middle of suburbia. But then again, in the Cybertruck, the line between the mundane and the extraordinary is as thin as the vehicle's paint job. Just remember, with great power comes great responsibility – and potentially a very embarrassing story to tell at parties.

## Learning the art of ejecting yourself from tedious talks about fuel efficiency.

In the fantastically unrealistic world of the Tesla Cybertruck, where the impossible is merely improbable, learning to eject yourself from boring conversations about fuel efficiency is an art form, a skill that every Cybertruck owner should master. Let's face it, discussing miles per kilowatt-hour is about as thrilling as watching paint dry, but with less color.

Now, the Cybertruck, being the futuristic behemoth it is, has a solution for everything, including social predicaments. Buried deep in the bowels of the owner's manual, between how to turn on your headlights and the correct way to admire the truck's angular beauty, lies the instructions for the ejector seat – Tesla's answer to social discomfort.

Using the ejector seat is simple in theory but requires the timing and precision of a Swiss watch. Imagine you're trapped in a conversation, cornered by the kind of person who thinks discussing the Cybertruck's battery range is appropriate party talk. You nod politely, your eyes glazing over, as they drone on about the virtues of electric propulsion. But fear not, for your escape is literally a button push away.

First, subtly maneuver yourself back into the driver's seat under the pretense of demonstrating the luxurious comfort of Tesla's vegan leather. Now, casually place your hand near the ejector button – a small, unassuming button labeled 'E.S.C.', which you assume stands for 'Escape Social Catastrophe'.

At the precise moment when your companion starts to expound on the merits of regenerative braking, you press the button. Instantly, the Cybertruck springs into action. The roof splits open with the grace of a blossoming flower, albeit one made of steel and glass. Then, with the force of a thousand unsuppressed sneezes, the ejector seat launches you skyward.

As you soar through the air, you experience a moment of serene tranquility, free from the shackles of mundane chatter. Below you, your conversation partner is left staring, mouth agape, possibly reconsidering their choice of small talk. The wind rushes past you, and for a brief moment, you're not just escaping a tedious conversation; you're an astronaut on a mission to explore strange new worlds, where people talk about interesting things at parties.

Of course, what goes up must come down. The parachute, cleverly concealed within the seat, deploys, allowing you to drift gently back to Earth. You land with the grace of a cat, albeit one that's just been shot out of a cannon. Your landing spot is crucial – aim for the buffet table for maximum impact and a soft landing. Sure, you'll ruin the hors d'oeuvres, but it's a small price to pay for freedom.

The key to mastering this swift escape is practice. You must be ready to deploy at a moment's notice, at the first sign of a conversation heading towards the kWh abyss. Timing, posture, and a good sense of wind direction are crucial. It's not just about getting out of the conversation; it's about doing it with style.

So, the next time you find yourself trapped in a soul-sucking discussion about the Cybertruck's fuel efficiency, remember the ejector seat. It's your secret weapon, your escape hatch, your ticket to freedom. Just be sure to apologize to the host for any disruptions caused – landing in the middle of a garden party with a deployed parachute tends to attract attention. But then again, owning a Cybertruck is all about making a statement, and what better statement than a dramatic aerial exit from the dullest of conversations?

## Use at Your Own Risk

In the fantastical, somewhat unhinged world of the Tesla Cybertruck, the inclusion of an ejector seat is less a feature and more a wild gamble. While the prospect of catapulting oneself out of tedious social situations or gridlocked traffic may seem appealing, it's important to remember that this feature comes with a disclaimer: Use at your own risk. And believe me, the risks are as numerous as the complaints from traditional truck owners about the Cybertruck's design.

Firstly, let's consider the mechanics of using the ejector seat. This manual, in a fit of uncharacteristic optimism, suggests that the seat provides a swift and elegant exit strategy. In reality, it's akin to strapping yourself to a firework. The act of launching oneself skyward in a vehicle not designed for flight is as ludicrous as trying to teach a goldfish to walk. The ejector seat may launch you out of an unwanted conversation about the merits of electric vehicles, but it's also likely to launch you straight into a world of legal and medical headaches.

Then there's the matter of landing. The Cybertruck's manual vaguely mentions a parachute for a "controlled descent." However, the chances of you mastering the parachute on your first impromptu flight are about as high as finding a vegetarian shark. The likelihood is that your descent will be as controlled as a shopping trolley on a steep hill. You may escape the clutches of a mundane conversation, but you'll likely find yourself plunging into the clutches of a nearby tree, or worse, someone's conservatory.

Let's not overlook the legal ramifications of using the ejector seat. The laws regarding the ejection of oneself from a moving vehicle are murky at best, but it's safe to assume that the local constabulary will take a dim view of it. You may find yourself explaining to a bemused officer why you decided to eject yourself over the neighborhood, a story that will sound about as believable as a politician's election promises.

And consider the social consequences. While ejecting yourself from a particularly dull barbecue might seem appealing, your dramatic exit will likely lead to a lifetime of social ostracism. You'll be known as the person who ruined Aunt Mabel's 80th birthday party with an unscheduled aerobatic display. Good luck getting invited to any event after that, unless it's a convention for amateur stunt drivers.

There's also the issue of precision. The ejector seat is not a precision instrument. It's a blunt tool, a one-size-fits-all solution to a problem that probably requires a more nuanced approach. You're aiming to escape a boring conversation, but you're just as likely to find yourself inadvertently gatecrashing a football match, a wedding, or even a funeral. The possibilities for unintended social disasters are endless.

In essence, the ejector seat of the Tesla Cybertruck is a classic case of 'just because you can, doesn't mean you should.' It's a feature that, on paper, promises excitement and adventure. In reality, it's a one-way ticket to a world of pain, legal troubles, and social pariah status. So, the next time you find yourself reaching for that ejector button, remember: the risks far outweigh the benefits. It's far safer, and more socially acceptable, to simply endure the conversation about fuel efficiency. After all, a few minutes of boredom is a small price to pay compared to a lifetime of being known as 'that person who ejected themselves from a barbecue.'

# 8. Driver "Aids" and Assumptions

Driver "Aids" and Assumptions in the Tesla Cybertruck are akin to having a backseat driver who not only has never driven a car but also thinks they're navigating a spaceship. These features, generously termed "aids," are about as helpful as a chocolate teapot and often come with assumptions that could lead to a comedy of errors – if they weren't so terrifying.

Let's start with the autopilot, Tesla's pièce de résistance. It's supposed to make driving as effortless as lounging on a beach in the Bahamas. In reality, it's more like trying to lounge on a beach during a hurricane. The system, with the confidence of a toddler doing a jigsaw puzzle, takes over the steering, lulling you into a false sense of security. You're supposed to keep your hands on the wheel, but the car behaves like an overbearing nanny, scolding you with alarms every time you attempt to make a manual correction. It's less of a co-pilot and more of a control freak.

Then there's the lane-keeping system, a feature that assumes roads are always perfectly marked and as straight as a Roman road. The reality, however, is that road markings are often as clear as the plot of a Christopher Nolan film. The Cybertruck, in its quest to keep you in the lane, occasionally takes creative liberties, veering with the grace of a drunken sailor. It's an adventure, but not the kind you signed up for.

The collision avoidance system sounds good on paper. It's like having a guardian angel who's constantly on the lookout for danger. But in practice, it's more like having a paranoid bodyguard. It sees threats everywhere – a paper bag on the road, a low-flying bird, a cloud that looks vaguely like a truck – and reacts with the subtlety of a sledgehammer. The car brakes so abruptly, you're left wondering if you've hit an invisible wall.

Adaptive cruise control is another marvel of modern technology that makes some bold assumptions. It assumes that all drivers behave predictably and that traffic flows as smoothly as a river. The truth is that driving among humans is more like trying to navigate a pinball machine. The Cybertruck, with its adaptive cruise control, often accelerates and brakes with the finesse of a learner driver, turning a relaxing drive into a herky-jerky rollercoaster ride.

The parking assist feature is supposedly there to help you park with the precision of a Swiss watch. Unfortunately, it's based on the assumption that all parking spaces are as generously sized as those in a Texas ranch. Attempt to squeeze into a normal-sized spot, and the Cybertruck's sensors light up like a Christmas tree, beeping frantically as if you're trying to dock the International Space Station.

And we mustn't forget the voice-controlled infotainment system. It promises to obey your every command, from changing the radio station to setting the navigation. However, it operates under the assumption that you speak in the Queen's English, enunciating every syllable perfectly. Speak with any sort of accent or colloquialism, and it descends into confusion, like a tourist with a phrasebook in a foreign land.

In summary, the driver "aids" in the Tesla Cybertruck assume a world of ideal conditions – perfect roads, rational drivers, and clear weather. The moment any real-world unpredictability is introduced, these systems get as confused as a cat at a dog show. They're not so much aids as they are overzealous electronic butlers, constantly trying to help but more often than not, spilling the soup instead. So, when you engage these features, be prepared for a ride that's as unpredictable as it is futuristic. It's like being in a sci-fi movie, but you're not sure if you're the hero or the extra about to be zapped into oblivion.

## Parking Assistance: The Bumper's Sacrifice

When it comes to parking the Tesla Cybertruck, the so-called parking assistance system is about as helpful as a chocolate fireguard. This system, which presumably was designed during a particularly whimsical moment in Tesla's engineering department, seems to operate under the assumption that both the vehicle and the surrounding infrastructure are made of rubber.

You see, in the hands of a sane individual, parking assistance should mean a gentle nudge, a helpful beep perhaps, or at most, a polite suggestion on when to turn the steering wheel. In the Cybertruck, it's more akin to a game of bumper cars at the fair. The system appears to be under the impression that as long as you're not catapulting pedestrians into the next postcode, it's doing a splendid job.

Let's paint a picture. You approach a parking space, the size of which would normally be considered adequate for a vehicle less resembling a small tank. The Cybertruck, with the confidence of a walrus performing a ballet, starts to take over the wheel. Now, any sane person would keep a cautious foot hovering over the brake pedal, but This manual, with its usual dose of over-optimism, assures you that the truck 'knows what it's doing'.

The sensors begin to beep, and at first, it's a comforting sound – a technological lullaby, lulling you into a false sense of security. But as the truck edges closer to the car behind, the beeping increases in urgency, like the heart rate of a squirrel that's just downed a double espresso. By the time you're mere inches away, it's a cacophony of alarms that could wake the dead.

But does the Cybertruck stop? Of course not. It gently nudges the car behind, as if testing the laws of physics. There's a sound – a crunch, a crumple, the kind of noise that makes your wallet weep. The parking assistance system, having sacrificed your bumper (and the other vehicle's) in the name of 'assistance', triumphantly declares that you are now parked.

The surrounding onlookers, who've been watching this spectacle with a mixture of horror and fascination, are now treated to the second act of the drama: your attempt to exit the parking spot. The Cybertruck, having already displayed its blatant disregard for the concept of space, now must extricate itself from the metal sandwich it has created.

The system, unfazed by the chaos it has wrought, cheerfully suggests reversing. You, now sweating like a tourist in a sauna, oblige. The beeping starts again, a relentless symphony that seems to be Tesla's way of saying,

"I told you so." With a grace that can only be described as elephantine, the Cybertruck lumbers out of the spot, leaving a trail of destruction in its wake.

And there you have it. You're out, the parking assistance system congratulating itself on a job well done, while you're left to contemplate the mysteries of the universe, such as why a vehicle equipped with the latest in automotive technology has the spatial awareness of a blindfolded hippopotamus.

In essence, the parking assistance in the Cybertruck is an adventure, a heart-pounding, palm-sweating escapade that will leave you questioning the very nature of help. It's not just a feature; it's a test of nerve, a challenge to your driving skills, and a blow to your faith in technology. So, the next time you engage the Cybertruck's parking assistance, remember: it's not just assisting you to park; it's sacrificing your bumper on the altar of progress.

## The Crunching Sound of Assistance

In the realm of the Tesla Cybertruck, the parking assistance system is akin to a dance partner with two left feet. It promises a tango, but what you get is more of a clumsy foxtrot, where stepping on toes is not just likely, it's almost guaranteed. This feature, festooned with sensors and cameras, appears to have been calibrated using the same principles as a bumper car ride.

When you first engage the parking assistance, there's a moment of naïve optimism. You expect a seamless ballet of automotive precision, but what unfolds is a performance so graceless it could only be appreciated at a demolition derby. The system beeps and flashes like a discotheque, creating an atmosphere of urgency that would make even the most Zen driver break into a sweat.

As the Cybertruck begins its approach to the parking space, it moves with the delicacy of a bull in a china shop. This manual, in a triumph of understatement, suggests you might hear 'minor contact' with surrounding objects. This is the equivalent of describing an elephant sitting on your foot as a 'slight pressure.' The first crunching sound you hear isn't just the Cybertruck acquainting itself with the neighboring vehicle; it's the sound of your insurance premium preparing for lift-off.

The theory behind this system is wonderful – a vehicle that parks itself, reducing driver stress and the risk of scrapes. However, the practice is akin to performing dental surgery with a sledgehammer – it's not just stressful, it's a spectacle of wincing and cringing. The Cybertruck, oblivious to your rising panic, continues its assault on the parking space, each beep a mocking reminder of your misplaced trust in technology.

As the vehicle judders to a halt, having finally 'parked', you're left to survey the scene. The parking space, which initially seemed ample, now looks like a crime scene. The Cybertruck, triumphant in its own mechanical mind, sits askew, looking like a guilty child surrounded by broken vases. The surrounding vehicles bear the scars of battle, and you can't help but feel a pang of sympathy for their unsuspecting owners.

Exiting the vehicle, you brace for the reactions of passersby, who've watched this parking fiasco unfold with a mixture of horror and amusement. You consider trying to explain that it was the parking assistance, not you, but realize it's a futile endeavor. It's like trying to convince someone that your dog ate your homework – technically possible, but wildly improbable.

The crunching sound of assistance, therefore, is a misnomer. It's not assistance; it's a mechanical interpretation of chaos theory. Each parking attempt is a roll of the dice, a gamble where the stakes are the bodywork of your vehicle and those unfortunate enough to be parked nearby.

So, the next time you consider using the Cybertruck's parking assistance, remember: it's less of an aid and more of an adventure, a foray into the unknown where the only certainty is the sound of metal on metal. It's an experience, a story to tell, though perhaps not one of triumph. Rather, it's a cautionary tale, a reminder that sometimes, the old ways – doing it yourself, with your own hands and eyes – are not just better, they're significantly less expensive.

## The Art of Guesswork: How to park using your instincts because the sensors are probably wrong.

Parking the Tesla Cybertruck is like playing a game of blind man's bluff, only the blindfold is a high-tech sensor system that's about as reliable as a weather forecast in Britain. In the Cybertruck, you quickly learn that the art of parking relies less on the slew of sensors and cameras adorning its metallic hide and more on good old-fashioned guesswork and a dollop of blind faith.

Let's face it, the sensors on the Cybertruck have the accuracy of a drunk throwing darts. They beep and flash with the enthusiasm of a child with a new toy, but the information they provide is about as useful as a chocolate teapot. You're informed of imminent collisions with obstacles that aren't there, and yet, when a real threat, like a bollard or a small car, lurks in the vicinity, the sensors remain as silent as a mime.

As you approach a parking space, the sensors begin their symphony of confusion. The screen lights up with a Christmas tree's worth of warnings, each less helpful than the last. You edge closer to the space, guided by the beeps, which increase in frequency and pitch until they resemble the soundtrack of a horror movie.

It's at this point you realize that you're better off relying on your own instincts. You recall the days of learning to park in a vehicle that offered nothing more than a rearview mirror and a prayer. You channel this primitive energy, turning off the sensor display, which by now looks like it's trying to communicate in Morse code.

You begin the delicate dance of parking. The Cybertruck, with its heft and angles, is about as easy to park as a small yacht. You twist and turn, peering out of the windows, making judgments and adjustments based on gut feeling rather than the panicked shrieks of the sensors.

The true challenge is in the spatial awareness – or the lack thereof. The Cybertruck feels as if it occupies a different dimension, where its proportions are a mystery even to itself. You edge back, using the force of your intuition to gauge the distance. The lack of sensor beeps is unnerving, but liberating. It's just you, the car, and the open space.

Parking becomes a meditative practice, a throwback to a simpler time when drivers relied on skill rather than technology. You find yourself becoming one with the Cybertruck, understanding its dimensions not through a screen or a series of frantic beeps, but through an innate sense of space and boundaries.

As you finally slot into the space, with mere centimeters to spare, you feel a sense of accomplishment. You've done it the old-fashioned way, using nothing but your wits and your instincts. You step out of the Cybertruck, glancing at the perfectly parked vehicle with a sense of pride.

Parking the Cybertruck, then, is an exercise in trust – not in the myriad of sensors and cameras, but in your own abilities. It's a reminder that sometimes, technology can promise the world but deliver a atlas. In a vehicle that feels like it's been beamed down from the future, it's ironic that the best way to park it is to hark back to the past, to rely on the art of guesswork, and to trust in your own human instincts. So next time you park your Cybertruck, remember: the sensors are more a source of entertainment than assistance. Trust yourself, and you'll likely find that you're the best parking sensor you've got.

## Hill Start: A Game of Chance

In the grand, bewildering world of the Tesla Cybertruck, performing a hill start is less a driving maneuver and more akin to spinning the roulette wheel in a casino – except here, the stakes are your sanity and the pristine state of your cybernetic behemoth.

Firstly, let's acknowledge the Cybertruck's hill start assist feature, which in theory, prevents the vehicle from rolling back on a steep incline. In practice, however, it's as reliable as a chocolate teapot. It promises to hold the vehicle steady, giving you time to move your foot from the brake to the accelerator. But the moment you take your foot off the brake, the truck lurches backward with all the enthusiasm of a lemming on a cliff edge. It's an instant where time seems to freeze, and you find yourself contemplating the meaning of life, the universe, and why on earth you thought driving this polygon on wheels up a hill was a good idea.

Then there's the accelerator pedal, which in the Cybertruck is more a suggestion than an actual command to the vehicle. On a hill start, you'd expect a gentle press to ease the truck forward. But no, the Cybertruck, with the overzealousness of a toddler at a birthday party, leaps forward as if it's been prodded with a cattle prod. The result is a jolt that can snap your head back faster than you can say, "I should have bought a Range Rover."

The regenerative braking system, another "aid" in this hill start fiasco, adds its own flavor of chaos to the mix. In theory, it's supposed to conserve energy by converting the kinetic energy from braking into electric power. In practice, on a hill, it behaves like a moody teenager. Sometimes it helps slow the vehicle down, sometimes it doesn't. It's as predictable as a British summer.

Let's not overlook the sheer size and weight of the Cybertruck. On a hill, these factors combine to create a sense of impending doom. The truck feels like it's not just fighting gravity, but engaged in an epic battle with the laws of physics themselves. As you attempt to coax it into moving forward, it's as if the entire vehicle is having an existential crisis, pondering whether it should go up, down, or just give up and become a very expensive roadside ornament.

This manual, in its usual style, glosses over these nuances with the breezy optimism of a salesperson who's never driven the truck up anything steeper than a speed bump. It suggests a smooth, controlled approach, but this is the Cybertruck – smooth and controlled are as foreign to it as a salad at a steakhouse.

Performing a hill start in the Cybertruck, therefore, becomes a game of chance. Each attempt is a roll of the dice, a flirtation with the twin specters of gravity and momentum. You're not just driving; you're gambling with a vehicle that has the power and weight of a small moon.

So, the next time you find yourself at the bottom of a steep hill in your Cybertruck, remember: this isn't just a hill start; it's an adventure, a challenge, a test of your mettle. Will you roll backward, lurch forward, or perhaps discover a hitherto unknown feature of the truck, like spontaneous flight? Only time, and your luck, will tell. But one thing is certain – it will be an experience, a story to tell, assuming, of course, you survive it. And in the world of the Cybertruck, that's really what it's all about – not just driving, but surviving the drive.

# Rolling the Dice: Will you move forward, roll back, or just stay put?

In the somewhat bizarre world of the Tesla Cybertruck, every attempt at movement resembles a game of chance, akin to rolling the dice in a back alley craps game. You, the driver, are not so much in control as you are a hopeful participant in a vehicular lottery where the odds are as predictable as a cat on a hot tin roof.

Take, for example, the simple act of pulling away from a stop. In any normal vehicle, this is a straightforward affair. In the Cybertruck, however, it's an exercise in clairvoyance. Will the truck surge forward like a startled rhino, or will it remain as motionless as a sullen teenager asked to clean their room? It's anyone's guess. You press the accelerator with the same trepidation as one might press a button labeled "Do Not Press" in a nuclear missile silo.

Then there's the question of whether you'll roll back. On a hill, this is a particularly poignant matter. You'd think, given all its technology, the Cybertruck would hold its position with the steadfastness of a guard outside Buckingham Palace. Instead, it's more likely to slowly creep backwards with all the stealth of a cat burglar. This would be fine if you were planning a discreet getaway, less so if there's a vehicle behind you whose driver is blissfully unaware of your impending reverse embrace.

Now, let's consider the scenario where the Cybertruck decides to just stay put. You're in a hurry, you have places to be, people to see, and yet the truck has decided that this is the perfect moment for a bit of meditation. No amount of cajoling, pedal pressing, or colorful language will persuade it. It's as if the truck is pondering the meaning of its existence, while you're left pondering the meaning of your car loan.

This manual, a document that seems to have been written in an alternate universe where physics is more of a guideline, suggests smooth, controlled inputs. This is akin to suggesting one should negotiate with a bear over the last piece of salmon. The Cybertruck's responses to these inputs range from apathetic to overly enthusiastic, with little regard for what lies in between.

As for stopping, that's another roll of the dice. The Cybertruck's brakes are about as predictable as a game of pin the tail on the donkey, after the participants have been spun around one too many times. Will you stop on a dime, or continue sailing forward as if the brakes were merely a decorative feature? The suspense is almost unbearable, a thriller played out in real-time every time you approach a red light.

In essence, driving the Cybertruck is less about transportation and more about participating in a game of chance. Every journey is a roll of the dice, a flirtation with the fickle whims of Tesla engineering. Will you move forward, roll back, or just stay put? The answer is as uncertain as the British weather.

So, the next time you climb into your Cybertruck, remember: you're not just driving a vehicle; you're entering a casino on wheels. The dice are in your hands, the odds are unclear, and the stakes are high. But one thing is certain – it's going to be an interesting ride.

## The Thrill of Uncertainty

When you're behind the wheel of a Tesla Cybertruck, embarking on a hill start is less about the mundane act of moving a vehicle forward and more about embarking on a thrilling escapade into the unknown. It's akin to playing a round of Russian roulette, only the gun is your Cybertruck and the bullets are the unpredictable outcomes of your hill start. Each time you put your foot on the accelerator, you're not just pressing a pedal; you're spinning the cylinder and pulling the trigger.

Firstly, there's the sheer anticipation. As you sit there, poised on an incline, the Cybertruck's array of sensors and computers are supposedly calculating the perfect amount of power needed to ascend gracefully. But, as with most things in the Cybertruck, what's supposed to happen and what actually happens are as closely related as the Queen and Ozzy Osbourne.

You press the accelerator, and that's when the fun begins. Will the Cybertruck leap forward like a startled gazelle, leaving a trail of rubber and bewildered pedestrians in its wake? Or will it roll backward, gently kissing the bumper of the car behind, in a metal-on-metal caress that's about as romantic as a slap with a wet fish?

The unpredictability is the spice of this hill start curry. It's a culinary concoction that could either be a delightful jalfrezi or a vindaloo that leaves you weeping for mercy. The Cybertruck might decide to do neither, instead choosing to sit there stubbornly, defying both gravity and your expectations, becoming a very expensive roadblock.

Let's not forget the regenerative braking system, which adds its own unique flavor to this recipe. Designed to recharge the battery when you brake, on a hill it sometimes has the enthusiasm of a teenager asked to do the dishes. It can't decide whether to help you hold your position or to release you into the wild, untamed laws of physics.

Then there's This manual's advice. It cheerily suggests a smooth application of the accelerator, a delicate balance of power and finesse. But when you're sitting in a Cybertruck, on a hill, with the laws of gravity baying for your dignity, finesse is about as useful as a chocolate teapot. You need the touch of a bomb defuser and the nerves of a tightrope walker.

The result is a hill start that's as much an adrenaline rush as bungee jumping, only without the safety harness. Your heart races, palms sweat, and you develop a newfound appreciation for flat, level ground. It's a rollercoaster of emotions, a test of your resolve, a trial by fire that leaves you either elated or wondering how to explain to your insurance company that gravity is a cruel mistress.

In essence, a hill start in the Tesla Cybertruck is not just a driving maneuver; it's an adventure, a leap into the unknown, a thrilling escapade on the razor's edge of automotive possibility. You're not just moving a vehicle from point A to point B; you're taking a gamble with a machine that has all the predictability of a game of chance.

So, the next time you find yourself facing the uphill challenge in your Cybertruck, remember: this isn't just driving, it's living on the edge. It's embracing the thrill of uncertainty, the adrenaline rush of the unknown. Will you move forward, roll back, or simply stay put? Only the Cybertruck knows, and it's not telling. But one thing is for sure – it's going to be an exciting ride.

## The Invisibility Cloak: Just Close Your Eyes

In the fantastical universe of the Tesla Cybertruck, there's a feature so advanced, so revolutionary, that it defies belief – the invisibility cloak. Yes, you heard that right. According to the pages of the Cybertruck's manual, which I'm convinced were written during a particularly vivid fever dream, this vehicle can render itself invisible. Now, before you get too excited and start planning your career as a superhero, let's explore what this really means.

First off, the so-called invisibility cloak is less Harry Potter and more Harry Houdini. It's an illusion, a trick of the light, a bit of technological sleight of hand that makes you think the truck can disappear. But in reality, it's about as effective at making the Cybertruck invisible as squinting really hard is at making your problems go away.

When you engage this feature, what happens? Does the Cybertruck vanish into thin air, leaving behind a puzzled crowd of onlookers? Does it blend into the landscape like a chameleon on a canvas of modern art? No. What happens is the screen on the dashboard lights up, displaying a graphic that suggests invisibility, while the truck remains as visible as an elephant in a living room.

This manual, in a masterpiece of overstatement, suggests using the invisibility cloak to avoid unwanted attention. Imagine, you're driving down the high street, and you activate the cloak. What happens? Do people gasp in amazement, wondering where the truck went? No, they continue sipping their lattes and walking their dogs, blissfully unaware of your attempts at subterfuge.

Then there's the small matter of actually driving while invisible. This manual, in its infinite wisdom, fails to mention the practicalities of navigating a two-tonne truck that you can't see. It's like trying to drive while wearing a blindfold, relying solely on hope and good intentions to avoid catastrophe.

The reality is that the invisibility cloak is just a fancy way of saying, "Just close your eyes and pretend." It's like being a child under a bedsheet, believing you're hidden from the world. The only thing disappearing is your grip on reality, as you sit there, eyes closed, in a truck that's about as invisible as a fireworks display on a dark night.

So, the next time you read about the Cybertruck's invisibility cloak, remember: it's not a doorway to an unseen world, it's a one-way ticket to disappointment. It's not a feature, it's a fantasy, a whimsical notion in a vehicle that's more grounded in science fiction than science fact.

In the end, the invisibility cloak is a metaphor for the Cybertruck itself – bold, audacious, and a little bit absurd. It promises the moon but delivers a handful of cheese. So, embrace the madness, enjoy the fantasy, but remember: when it comes to being invisible, you're better off sticking to hiding behind the sofa. At least there, you won't risk a multi-car pile-up on the motorway.

## Pretending your car can become invisible, because why not?

In the mystifying, logic-defying world of the Tesla Cybertruck, the power of imagination isn't just encouraged; it's practically a requirement. Among the most bewildering of its supposed features is the ability to turn invisible. Yes, you heard that right. The Cybertruck, a vehicle more conspicuous than a streaker at a cricket match, apparently has the ability to disappear like a shy ghost. While this may sound as plausible as a chocolate radiator, it's crucial for Cybertruck owners to have a spirited sense of imagination.

Let's delve into the realm of make-believe, where your Cybertruck can become invisible at the push of a button. As you press this magical button, nothing happens. Absolutely nothing. But that's where the power of imagination comes into play. You must believe. Close your eyes (figuratively, not literally, especially if you're driving), and transport yourself into a world where your giant metal trapezoid can suddenly become as elusive as the concept of a quiet politician.

Picture the scene: you're stuck in traffic, surrounded by a sea of mundane, visible cars. You tap the invisibility button, and voilà – in your mind, you have vanished. The other drivers can't see you because, in your rich tapestry of imagination, you're as invisible as a decent plot in a reality TV show. Never mind that in reality, you're causing quite the spectacle as drivers around you wonder why you're grinning maniacally in a truck that looks like it was designed on a Etch A Sketch.

Think of the practical applications of an invisible vehicle. You could avoid traffic wardens, dodge bothersome in-laws, or simply enjoy the baffled looks of pedestrians as they try to rationalize why a floating driver is cruising down the high street. In this fantastic world, the Cybertruck is not just a means of transportation; it's a cloak of invisibility, a chariot of stealth, a vessel of incognito travel.

But back in the real world, we must face the music, or rather, face the fact that the Cybertruck is as invisible as an elephant in a ball pit. The idea that this hulking piece of futuristic bravado could blend into its surroundings is as likely as finding a needle in a haystack – a haystack made entirely of needles.

So, why include such a preposterous feature in This manual? Because, in the world of the Cybertruck, reality is not a limitation, it's an inconvenience. This manual isn't just a guide; it's a portal to a parallel universe where the laws of physics are mere guidelines, and the impossible is merely improbable.

Embracing the power of imagination with the Cybertruck allows you to transcend the mundane, the everyday, the routine. Why settle for being stuck in traffic when you can be stealthily lurking in plain sight? Why be just another vehicle on the road when you can be an unseen specter of automotive mystery?

In summary, the invisible Cybertruck is less about actual stealth technology and more about the stealth of your imagination. It's about believing in the unbelievable, embracing the absurd, and accepting that in the world of Tesla, anything is possible, even when it's patently not. So go ahead, press that invisibility button, and disappear into a world of fantasy. Just remember to reappear before you reach your destination. After all, it's hard to explain to the valet why they need to park a car they can't see.

## Exploring the exciting possibility of not seeing things right in front of you.

In the fantastical world of the Tesla Cybertruck, the blind spot isn't just a minor inconvenience – it's an entire event, a bonanza of not seeing things that are blatantly obvious to everyone but you. The Cybertruck, with its design straight out of a dystopian sci-fi movie where everyone is apparently too cool to look around properly, has taken the concept of a blind spot and elevated it to an art form.

Let's get this straight: the blind spot in most cars is like that small, annoying bit of the windscreen that the wipers can't quite reach. In the Cybertruck, however, it's more akin to wearing a pair of horse blinkers while trying to admire the panoramic view at the Grand Canyon. It's not so much a blind spot as it is a blind continent.

Now, you might think that with all the cameras, sensors, and gizmos plastered all over the Cybertruck like barnacles on a ship's hull, blind spots would be a thing of the past. Ah, but that's where the fun begins. The

Cybertruck treats visibility like it's an optional extra, a luxury that only the weak rely on. Why simply look and see when you can embark on a thrilling game of automotive hide-and-seek?

Imagine you're cruising down the motorway, the sun is shining, and you decide it's time to change lanes. In any normal car, a quick glance over your shoulder suffices. But this is the Cybertruck – you don't simply glance over your shoulder. No, you engage in a complex ritual involving multiple camera feeds, predictive algorithms, and a quick prayer to the gods of blind luck. You signal, you turn the wheel, and then... surprise! You're greeted with the heart-stopping excitement of discovering a vehicle in your blind spot that was previously as detectable as a stealth fighter jet.

This manual, bless its optimistic soul, tries to paint this as a feature. It suggests that the blind spot is there to keep drivers on their toes, to add a bit of excitement to the mundane task of driving. It's like a game show, but instead of winning a prize, you get to test the limits of your vehicle's safety features and the strength of your own heart.

And let's not overlook the rearview mirror – or, as I like to call it in the Cybertruck, the 'guess-what's-behind-you mirror'. With the rear window offering all the visibility of a medieval arrow slit, you're forced to rely on a screen displaying a feed from a camera that seems to be more interested in the sky than the road behind you.

In summary, driving the Cybertruck with its blind spot bonanza is not just a journey from A to B; it's an adrenaline-fueled voyage into the unknown. It's a reminder that, in the world of Tesla, what you can't see can indeed hurt you, but discovering it is half the fun. So, embrace the uncertainty, cherish the heart palpitations, and remember – in the Cybertruck, every lane change is a roll of the dice, every rear glance an exploration into the exciting world of the unseen. It's not just driving; it's an adventure in trust, technology, and the thrillingly unexpected.

# 9. Vehicle Neglect

In the realm of the Tesla Cybertruck, the notion of vehicle neglect takes on a whole new meaning. It's like having a pet rock: you assume it requires no maintenance, and one day you find out it's actually a very expensive, very stationary boulder. The Cybertruck, with its austere, post-apocalyptic aesthetic, gives the misleading impression of being as low maintenance as a cactus in the desert. However, in reality, it's more like a diva on the red carpet, requiring constant attention and pampering.

Let's start with the basics: cleaning. You might think the Cybertruck, with its angular, brutish design, can withstand the slings and arrows of outrageous fortune, not to mention a bit of dirt. But no, neglect to clean it, and it starts to look less like a futuristic vehicle and more like a prop from a low-budget dystopian movie. Dirt and grime don't just stick to the Cybertruck; they cling to it like gossip in a small town. And trying to clean it? That's an endeavor akin to giving a cat a bath.

Then there's the battery, the heart and soul of the Cybertruck. Ignore the battery, and it will repay the favor by leaving you stranded, as powerless as a mobile phone on 1% battery at a music festival. This manual, in its infinite wisdom, suggests that the battery requires about as much attention as a pet rock. But just like with pet rocks, one day you find out there's more to it. Battery neglect in the Cybertruck doesn't just lead to reduced performance; it leads to a vehicle as mobile as a beached whale.

What about tire maintenance? The Cybertruck's tires are so large, they could be used as a makeshift home if you ever find yourself homeless (a distinct possibility if you spend all your money on the truck). Neglect to check the tire pressure, and you'll find that the Cybertruck handles corners with all the grace of a drunken elephant on roller skates. The tires don't just whisper their discomfort; they scream it, like an opera singer with a stubbed toe.

The suspension system, too, demands your attention. Ignore it, and the ride becomes less like gliding on air and more like riding a mechanical bull. Every bump in the road is felt, every imperfection magnified until you're convinced you're driving over a landscape of boulders.

And let's not forget about software updates. These are to the Cybertruck what spinach is to Popeye. Ignore them, and the truck's performance doesn't just dip; it plummets faster than the value of a used sports car. Each missed update turns the Cybertruck into a sulky teenager, refusing to do what it's told, its myriad of features and functions becoming about as cooperative as a cat in a bath.

In essence, neglecting the Cybertruck is a surefire way to turn this marvel of modern engineering into a very expensive lawn ornament. It requires a level of care and attention that's inversely proportional to its rugged appearance. The Cybertruck may look like it can survive an apocalypse, but treat it with the same cavalier attitude as one might have towards vehicle maintenance in a Mad Max film, and it'll be about as useful as a chocolate teapot. So, remember: treat the Cybertruck with the care and respect it demands, or be prepared for a future where your most significant relationship is with the tow truck driver.

# The Myth of Scratch Removal

In the world of the Tesla Cybertruck, the concept of removing scratches is as mythical as the Loch Ness Monster – often talked about, but never actually witnessed. You see, the Cybertruck, with its stainless steel exoskeleton, promises a future free from the heartache of scratches and dings. But much like expecting a quiet day on Twitter, this is a hope that is destined to be dashed.

Let's start with the stainless steel body – a material chosen, presumably, because someone thought it would be funny to make a truck that can double as a mirror for low-flying aircraft. This shiny surface is as impervious to scratches as a marshmallow is to a sledgehammer. This manual may claim that the truck is resistant to minor scrapes, but in reality, even a hostile glare seems to leave a mark.

Now, you might think that removing these blemishes would be a simple affair. A bit of polish, a soft cloth, perhaps a gentle touch. Oh, how delightfully wrong you would be. Trying to remove a scratch from the Cybertruck is like trying to remove a stain from a politician's reputation. No matter how hard you try, you're just going to make it more noticeable.

This manual suggests using a special cleaning compound, which is about as effective as using salad dressing. You apply it with hope in your heart, and yet, the scratch remains, as stubborn as a cat that refuses to move off your laptop keyboard. It's not just there; it's now shiny and more noticeable, proudly displaying your failure to every passerby.

But wait, This manual has more advice. It speaks of professional services, of experts who can make your Cybertruck look as good as new. So, you take it to these so-called experts, full of optimism. Hours later, you return, only to find that your once-pristine truck now looks like it's survived a sandstorm. The scratches are still there, but now they've brought friends.

The truth is, scratch removal in the Cybertruck is a myth, a fairy tale, a story we tell ourselves to feel better about the inevitable. It's like trying to smooth out wrinkles on an elephant – a noble effort, but ultimately futile. The Cybertruck, for all its futuristic bluster, handles scratches with all the grace of a toddler with a crayon.

In essence, owning a Cybertruck and trying to keep it scratch-free is like trying to keep a white shirt clean at a spaghetti-eating contest. It's not a question of if it will get scratched, but when. And when it does, you'll come to understand that these marks are not flaws. They're character, they're stories, they're battle scars from a world that's not quite ready for a vehicle that looks like it was designed for a Mars mission.

So, embrace the scratches, cherish the dings. Each one is a testament to the life of the Cybertruck, a vehicle that's more about making a statement than about blending in. After all, in a truck that looks like it could survive an apocalypse, what's a little scratch? It's not a blemish; it's a badge of honor.

## Scratches: A Badge of Honor

In the peculiar and often perplexing world of the Tesla Cybertruck, owning a vehicle without scratches and dents is like trying to find a quiet spot in a rock concert. It's an exercise in futility. This is especially true in the Cybertruck, a vehicle that attracts scratches and dents with the same enthusiasm as a magnet attracts iron filings.

Let's face it, the moment you take your shiny, new Cybertruck out for a spin, it's about as pristine as a dinner plate at a toddler's birthday party. The stainless steel exterior, which Elon Musk probably claimed could withstand a minor nuclear explosion, is actually more prone to scratches than your ego on a bad hair day.

Now, you might initially approach these blemishes with the same horror as finding a scratch on your favorite watch. But, in the grand tradition of Cybertruck ownership, you'll quickly learn to embrace these imperfections. After all, in the world of the Cybertruck, scratches and dents aren't flaws; they're stories, tales of adventure, of narrow misses, of that time you tried to navigate through a narrow alley and misjudged the width by a mere foot or two.

Each scratch on the Cybertruck is a badge of honor. It's a testament to its use, a symbol of your bravery in taking this behemoth on wheels into the wild urban jungle. The Cybertruck, with its post-apocalyptic design, isn't meant to be a garage queen. It's a warrior, a vehicle that wears its battle scars with pride.

This manual, while coy about this, subtly implies that scratches add character. In a section usually reserved for cleaning advice, there's a whimsical note suggesting that a well-lived Cybertruck is one that bears the marks of its experiences. It's like a face with laugh lines – sure, it's not as smooth as it once was, but it's infinitely more interesting.

And let's talk about the dents. In a normal car, a dent is a disaster, a calamity, a reason to call your insurance company. In the Cybertruck, it's a conversation starter. Each dent has its own story, its own unique origin. Was it from that rogue shopping cart that came out of nowhere? Or perhaps from that low-hanging branch you didn't see while admiring the Cybertruck's reflection in a shop window?

You'll soon find that these scratches and dents become part of your vehicle's unique character. They transform the Cybertruck from a mere vehicle into a trusted companion, a reliable steed that's been through the wars with you. It's no longer just a mode of transport; it's a repository of memories, a metallic diary on wheels.

So, when you see a new scratch or dent on your Cybertruck, don't despair. Embrace it. Celebrate it. Share its story with fellow Cybertruck enthusiasts, who will no doubt regale you with tales of their own vehicular battle scars.

In conclusion, the Cybertruck isn't just a vehicle; it's a lifestyle, a statement. It's a rugged, resilient machine that's designed to take on the world, scratches, dents, and all. These aren't imperfections; they're the marks of a life well-lived. They're not just scratches; they're stripes earned in the urban jungle. They're not just dents; they're anecdotes in metal form. In the world of the Cybertruck, every scratch and dent is a badge of honor, a sign that you're using the truck as intended – boldly, bravely, and unapologetically.

# The Futility of Buffing: Why trying to remove scratches might just give you more.

In the adventurous life of a Tesla Cybertruck owner, attempting to buff out scratches is akin to trying to solve a Rubik's Cube while blindfolded – a task fraught with frustration, futility, and the ever-looming prospect of making things much, much worse. The Cybertruck, with its gleaming, stainless steel exterior, presents itself as an indestructible beast from the future, but in reality, its skin is as delicate as the ego of a Hollywood starlet.

Firstly, let's understand the material we're dealing with here. Stainless steel might sound as robust as Superman on steroids, but in terms of resisting scratches, it's more like Superman after a Kryptonite sandwich. The Cybertruck's body, when exposed to the rigors of the real world – which includes anything from a rogue shopping trolley to an overzealous cat – collects scratches like a politician collects empty promises.

Now, upon discovering a scratch, the natural instinct of any proud vehicle owner is to try and buff it out. In the case of the Cybertruck, this is where the descent into madness begins. You see, buffing stainless steel requires the precision of a Swiss watchmaker and the gentle touch of a brain surgeon. The average Joe, armed with a buffing cloth and a can-do attitude, is more likely to add to the artwork rather than restore the canvas.

This manual, in a stroke of comedic genius, suggests a variety of products and techniques to tackle these blemishes. These range from specialist compounds, which smell like a chemical plant accident, to mysterious polishes that seem to have been concocted by a medieval alchemist. You apply these potions with hope in your heart, only to find that not only is the scratch still visible, but its surrounding area now resembles a miniature reenactment of the surface of the moon.

Then there's the use of buffing tools. This manual casually mentions the use of these with the nonchalance of recommending a teaspoon for a cup of tea. In reality, using a buffing tool on the Cybertruck is like using a chainsaw for a manicure. The tool, supposedly your ally in the war against scratches, becomes an agent of chaos, indiscriminately adding swirls and marks with the enthusiasm of a toddler in a sandbox.

And let's not forget the psychological toll this task takes. What starts as a simple mission to remove a scratch evolves into an obsession. The more you buff, the more scratches appear, and the more your sanity starts to fray at the edges. It's a Sisyphean task, where each attempt to improve the situation only serves to compound it.

In the end, trying to remove scratches from the Cybertruck is an exercise in futility. It's a battle against the forces of nature, a clash against the very essence of what it means to drive a vehicle as audacious and as ostentatious as the Cybertruck. These scratches, rather than being seen as flaws, should be embraced as part of the vehicle's character. They are not blemishes; they are badges of honor, each one telling a story of adventures had and challenges faced.

So, the next time you notice a new scratch on your Cybertruck, resist the urge to reach for the buffing cloth. Instead, step back, admire the mark for what it is – a symbol of a life lived to the fullest – and remember that in the grand scheme of things, it's the experiences, not the appearance, that truly matter. In the world of the Cybertruck, scratches are not just inevitable; they are a testament to the vehicle's use and enjoyment. Embrace them, and in doing so, embrace the true spirit of what it means to own such a unique piece of automotive history.

# The Eternal Quest for a Clean Windscreen

In the bizarre and often maddening world of the Tesla Cybertruck, keeping the windscreen clean is akin to trying to keep a white shirt spotless at a spaghetti-eating contest. It's an eternal quest, a never-ending battle against the forces of nature, road grime, and the inexplicable design choices of Elon Musk and his merry band of pranksters at Tesla.

Firstly, let's consider the sheer size of the Cybertruck's windscreen. It's massive, a panoramic vista that seems to have been designed with the idea that drivers might want to simultaneously drive and bird-watch. The problem is, with great glass comes great responsibility – the responsibility of keeping it clean. And in this regard, the Cybertruck is about as cooperative as a cat being bathed.

The wipers, for instance, seem to have been designed by someone who thought they were creating an abstract art installation. They swish back and forth with all the effectiveness of a pair of wet noodles. When confronted with anything more challenging than a light mist, they smear and smudge with the enthusiasm of a toddler armed with a crayon. You'd have better luck trying to clean the windscreen with a loaf of bread.

Then there's the windscreen washer fluid. In a normal car, this is a simple affair – a squirt of soapy water, and you're done. But in the Cybertruck, it's as if the washer fluid is mixed with a special concoction of grease and despair. It leaves streaks, smears, and occasionally, an existential crisis in its wake. You press the washer button in hope, only to watch in horror as your visibility is reduced to that of a mole in a coal mine.

Let's not forget about the dust and dirt. The Cybertruck, with its apocalyptic styling, seems to attract grime like a magnet. A drive through the countryside leaves you with a bug collection worthy of a natural history museum. A jaunt through the city, and the windscreen is coated with a film of urban detritus that's as hard to remove as gum from hair.

This manual, in a display of optimism that borders on delusion, suggests regular cleaning with a soft cloth and a gentle cleaner. This is about as effective as trying to stop a charging bull with a feather. You scrub, you wipe, you exert more effort than should be legally required, and still, the grime clings on, as stubborn as the Cybertruck's fan base.

And then there's the task of cleaning the inside of the windscreen – a task that requires the flexibility of a gymnast and the patience of a saint. The angle of the dashboard, combined with the expanse of the glass, means reaching every corner is a feat of human contortion. You find yourself in positions that would make a yoga instructor weep, all in the pursuit of clarity.

In essence, the quest for a clean windscreen in the Cybertruck is a Sisyphean task. It's a never-ending battle against the elements, against the inadequacies of wiper technology, and against the very laws of physics. But it's a battle that must be fought, for the windscreen is your window to the world, your portal to the road ahead.

So, when you embark on this noble quest, arm yourself with patience, with fortitude, and perhaps a small army of cleaning supplies. Remember, in the world of the Cybertruck, a clean windscreen is not just a luxury; it's a triumph, a victory against the odds, a testament to your perseverance and dedication. Embrace the challenge, and when you do achieve that fleeting moment of clarity, bask in the glory of your achievement – for it is fleeting, and soon you'll be back at it, wipers flailing, cloth in hand, fighting the good fight in the eternal quest for a clean windscreen.

## Mastering the art of achieving a streak-free windscreen, or at least trying to.

In the grand, bewildering scheme of owning a Tesla Cybertruck, the pursuit of a streak-free windscreen is much like trying to teach a cat to fetch – a venture filled with hope, despair, and a creeping realization that you're engaging in the impossible. The windscreen of the Cybertruck, vast as the Serengeti and just as challenging to navigate, seems to have been designed with the sole purpose of defeating any attempt at achieving optical clarity.

Firstly, let's address the wipers – or as I like to call them, the smudgers. They appear to have been designed by someone who misunderstood their purpose as being to redistribute dirt in a uniform pattern across the glass. They swipe with all the effectiveness of a bank card that's been demagnetized. Each pass across the windscreen leaves behind a trail of streaks and smears, a tangible record of their inadequacy.

Then there's the washing fluid, a liquid that's presumably meant to cleanse but instead adds its own unique layer of filth to the mix. It's as if someone filled the reservoir with a concoction of dish soap and despair. Activating the washers is akin to setting off a miniature foam party on your bonnet – fun for a moment, but leaving you with nothing but a mess.

Of course, This manual, in its eternal optimism, suggests a regular cleaning regimen with a soft cloth and specialized cleaner. This is akin to suggesting a gentle pat on the back as a remedy for being trampled by a herd of wildebeest. You set about the task with a cloth, armed with the enthusiasm of a soldier going into battle, only to realize that you are woefully ill-equipped. The sheer expanse of the windscreen, combined with the stubbornness of the streaks, makes cleaning it feel like an attempt to polish the floor of Grand Central Station with a toothbrush.

The interior of the windscreen presents its own special circle of hell. Attempting to reach the lower recesses of the glass requires the flexibility of an Olympic gymnast and the patience of a saint. You contort yourself into positions that defy anatomy, all the while battling the streaks that refuse to yield. The end result is a patchwork of smudges that give the impression of driving through a perpetual fog.

And let's not forget the environmental factors. The Cybertruck seems to attract dirt, dust, and grime with the magnetic force of a black hole. A mere drive down the street can leave the windscreen looking like it's been through a Saharan sandstorm. The elements conspire against you, leaving their mark with every raindrop, every speck of dust, every unfortunate insect that meets its demise on your glassy façade.

In essence, achieving a streak-free windscreen in the Cybertruck is a task of Herculean proportions. It's a battle against the elements, against the inherent design flaws of the vehicle, and against the very laws of physics. But it's a battle worth fighting. For those rare, fleeting moments when you do achieve clarity, the world seems a brighter place. You can bask in the glory of a job well done, even if it's only a matter of time before the cycle begins anew.

So, when you embark on this quixotic quest for a clear view, remember that it's about more than just cleaning a windscreen. It's about perseverance in the face of adversity. It's about the triumph of hope over experience. It's about mastering the art of the impossible. And when all else fails, it's about knowing when to give up, sit back, and enjoy the streaky view. After all, in the grand, perplexing world of the Cybertruck, sometimes the journey is more important than the clarity of the view ahead.

## The ongoing battle between your wipers and the elements.

In the curious and often frustrating world of the Tesla Cybertruck, the windshield wipers engage in a never-ending, Sisyphean struggle against the elements, a battle as lopsided as a one-legged man in a backside-kicking contest. These wipers, presumably designed by someone who thought rain was just a myth, are about as effective as using a sieve to bail water out of a sinking boat.

Let's start with the design. In a vehicle that looks like it was designed using a ruler and a set of children's building blocks, the wipers resemble an afterthought, like someone at Tesla suddenly remembered, "Oh, it might rain on Mars." They are the David to the Goliath that is the Cybertruck's vast windshield, comically undersized and about as effective as a chocolate teapot.

When you first turn them on, there's a moment of hope. Maybe, just maybe, they'll clear your view. But that hope quickly fades, replaced by the dawning realization that you're witnessing a masterclass in futility. The wipers swipe with all the enthusiasm of a teenager asked to clean their room. They leave behind streaks and smears, a mosaic of grime that only serves to make your view of the road ahead more akin to looking through a frosted glass window.

And then there's the speed. The wipers have two settings: too slow to be useful, and so fast they might fly off and enter orbit. The first setting is like watching paint dry, while the second is like being in the front row of a drum solo. Neither is particularly helpful in an actual downpour, when what you need is a setting that's just right, not a binary choice between a tortoise and a hare on steroids.

But the real fun begins when you encounter anything other than water. Mud? The wipers smear it around like an abstract expressionist. Snow? They pack it down until you have a miniature ski slope on your windshield. Leaves? They turn into a salad spinner, flinging bits of foliage all over with wild abandon.

This manual, with its characteristic overconfidence, suggests that the wipers are part of the Cybertruck's sophisticated environmental response system. This is akin to saying a paper umbrella is suitable for a hurricane. In reality, the wipers are about as sophisticated as a pair of rubber bands attached to a motor.

Then there's the issue of maintenance. The wipers seem to wear out faster than a politician's promise. You replace them, hoping for an improvement, only to find that they degrade from 'barely adequate' to 'downright dangerous' in the time it takes to say, "I should have bought a Land Rover."

In conclusion, the ongoing battle between the Cybertruck's wipers and the elements is less a battle and more a one-sided rout. The elements have the upper hand, always one step ahead, leaving the wipers to flail and thrash in a pitiful display of ineptitude. It's a war that can't be won, a fight that can't be conquered. All you can do is sit back, turn on the wipers, and watch the tragic-comic ballet of rubber and rain play out on the vast stage of your windshield. In the world of the Cybertruck, the wipers aren't a tool; they're entertainment, a slapstick performance that provides amusement, if not clarity.

## Anti-Rust Coating: Also Known As Mud

In the daring and often ludicrous world of the Tesla Cybertruck, the concept of anti-rust coating takes on a whole new, rather earthy dimension. Here, in the land of the Cybertruck, where the usual laws of automotive care are as malleable as a politician's promise, the conventional anti-rust coatings are about as useful as a chocolate teapot. Instead, the Cybertruck introduces a novel, revolutionary method to combat rust – a generous coating of mud.

Let's face it, the Cybertruck, with its stainless steel exoskeleton, was supposedly as impervious to rust as a duck is to water. However, in the real world, where roads are awash with more salt than a fast-food meal and more grime than a teenager's bedroom, this shiny exterior is tested to its limits. The solution, as Cybertruck owners have quickly discovered, isn't found in some high-tech, polymer-based, nanotechnology-infused coating. No, it's found in a good, old-fashioned layer of mud.

This approach, which may seem as backwards as a screen door on a submarine, is actually steeped in logic – a sort of back-to-basics, nature's-own-protectant kind of logic. The mud, generously applied by the simple act of driving on any road that isn't as smooth as a billiard table, forms a protective barrier. It's like a suit of armor, albeit one that's made from earth and looks like something a child would make in pottery class.

Now, the application of this anti-rust coating, or mud, as we laypeople call it, is a masterclass in precision engineering. It requires just the right mixture of water and soil, applied at just the right speed. Too little water, and you're just dusting your truck in a fine silt. Too much water, and you're giving it a mud bath. But get it just right, and voilà – you have a protective coating that's as effective as it is rustic.

The beauty of this method, aside from giving your Cybertruck a rugged, just-off-the-Safari look, is that it's self-renewing. Each journey adds another layer, enhancing the protective qualities and ensuring that the stainless steel underneath remains as untouched as a salad in a steakhouse.

This manual, in its typical style, glosses over this unconventional anti-rust method with the breezy nonchalance of a British aristocrat discussing the weather. It hints at the 'self-preserving qualities of natural elements' and 'the vehicle's symbiotic relationship with the environment.' This is a fancy way of saying, "Don't worry if it gets dirty; it's supposed to."

And let's not forget the added benefits of this mud coating. It's a theft deterrent – who wants to steal a vehicle that looks like it's been excavated from an archaeological dig? It's a conversation starter – nothing brings people together quite like discussing the latest layer of muck on your vehicle. And it's a statement – in a world of polished, pristine vehicles, the mud-covered Cybertruck stands out like a sore thumb, a badge of honor that screams, "I use my truck, and I'm not afraid to show it."

In essence, the anti-rust coating of mud on the Cybertruck isn't just a practical solution; it's a lifestyle choice. It's embracing the messiness of adventure, the joy of the journey, and the satisfaction of knowing that beneath the layers of earth and grime, your Cybertruck is preserved in all its stainless steel glory. So, the next time you look at your mud-caked Cybertruck, remember: that's not dirt; it's a protective coating, a mark of distinction, a shield against the ravages of time and the elements. It's not just mud; it's the future of anti-rust technology, as envisioned by Tesla.

## Understanding how a layer of mud can be just as effective as any anti-rust coating.

In the intriguing and often baffling world of the Tesla Cybertruck, the concept of rust protection takes a detour into the realm of the absurd. Here, in the land where normal rules are about as relevant as a chocolate teapot, the traditional anti-rust coating is replaced with something far more primitive, yet unexpectedly effective – a good, thick layer of mud. Yes, mud. That stuff you spend your childhood playing in and your adult life trying to avoid.

Firstly, let's be clear about one thing: when Elon Musk was designing the Cybertruck, it's as if he looked at traditional car care practices and thought, "No, let's not do that." So, instead of the usual chemical concoctions that most car manufacturers recommend for rust prevention, Cybertruck owners are inadvertently turning to nature's very own protector – mud.

Now, some might say that allowing your shiny, futuristic Cybertruck to get covered in mud is like letting a three-year-old decorate a wedding cake. But, as any Cybertruck owner will tell you, once you get past the initial shock of seeing your vehicle looking like it's just competed in a Tough Mudder challenge, there's something oddly reassuring about that layer of muck.

This isn't just any old dirt. This is a coat of armor, handcrafted by Mother Nature herself. As it turns out, a layer of dried mud is surprisingly effective at keeping the elements at bay. Water, salt, and other rust-promoting substances find it difficult to penetrate this organic shield. It's like having an army of tiny soldiers guarding every inch of your vehicle's body, except these soldiers are made of earth and grime.

The application process for this natural anti-rust coating is remarkably simple: drive your Cybertruck. That's it. No fancy equipment needed, no expensive products. Just the great outdoors and a sense of adventure. Every puddle splashed through, every dirt road traversed, adds another layer to your vehicle's protective barrier.

Of course, there are those who might question the aesthetics of this approach. To them, a car covered in mud is like a pair of shoes covered in dog muck. But what they fail to understand is the sheer joy that comes from knowing you're protecting your vehicle in the most natural way possible. Plus, there's a certain rugged charm to a mud-covered Cybertruck, like a rhino wearing its battle scars with pride.

This manual, ever the beacon of optimism, delicately refers to this as the 'benefits of natural coatings' and suggests that owners embrace the 'organic aesthetic.' This is a nice way of saying that your Cybertruck will often look like it's been parked in a swamp but fear not, it's all for the greater good.

And let's not forget the added benefits. A mud-covered Cybertruck is less likely to be stolen – most thieves prefer their getaway cars a little less conspicuous. It also makes for a great conversation starter at traffic lights and petrol stations. "Yes, it's supposed to look like that," you'll find yourself saying, followed by a knowing nod that suggests you're part of an exclusive club that understands the true meaning of vehicle protection.

In summary, the natural protector, the layer of mud on your Cybertruck, is not just a barrier against rust; it's a statement. It says that you're not just a driver; you're an explorer, an adventurer, a pioneer in automotive care who doesn't shy away from a bit of dirt. So, the next time you see your Cybertruck coated in a layer of the earth's finest, remember that this isn't just mud. It's a badge of honor, a testament to your vehicle's resilience, and proof that sometimes, the best solutions are the simplest. In the world of the Cybertruck, mud isn't just mud – it's a natural protector, a guardian against the ravages of time and the elements.

## Embrace the Dirt

In the world of the Tesla Cybertruck, cleanliness is not next to godliness. It's more akin to an unattainable fantasy, like a peaceful family Christmas dinner. Here, in this brave new world of stainless steel and electric motors, being perpetually dirty isn't just a likelihood; it's practically a design feature. The Cybertruck, with its rugged, post-apocalyptic aesthetic, seems to invite dirt, grime, and all manner of filth with open arms, like a long-lost friend.

Now, for the uninitiated, this may seem counterintuitive. After all, isn't the whole point of owning a vehicle like the Cybertruck to show it off, gleaming and pristine, as you cruise down the high street? Well, yes and no. You see, in the case of the Cybertruck, dirt doesn't just happen; it's part of the experience. It's the vehicular equivalent of a well-worn leather jacket, each stain and mark a story, a memory, a badge of honor.

Let's start with the basics. The Cybertruck, in its natural habitat, is as likely to remain clean as a toddler in a mud pit. The moment you take it off-road, which let's face it, is the whole point of owning a truck that looks like it could survive an interplanetary expedition, it starts collecting dirt like a politician collects scandals. Mud splashes up the sides, dust coats the windows, and before you know it, your shiny electric behemoth looks like it's just returned from a rally in the Sahara.

But here's where you need to adjust your perspective. This isn't just dirt; it's protective camouflage. In the wild, animals use mud as a sunscreen, a bug repellent, a cooling agent. The Cybertruck uses it as a shield against the elements. The layer of grime protects against scratches from rogue branches, chips from flying stones, and the envious glares of those who still think cars should be shiny.

And let's talk about the aesthetics. A clean Cybertruck, while undeniably beautiful, is also slightly intimidating, like a perfectly groomed show dog. But a dirty Cybertruck? That's approachable, it's relatable. It has character. It tells the world that you're not afraid to get your hands (and your vehicle) dirty. You're an adventurer, a trailblazer, a pioneer. You're not just driving a truck; you're making a statement.

Of course, This manual, in its usual optimistic manner, glosses over this. It offers token advice on cleaning, but deep down, even it knows the truth. The Cybertruck isn't just a vehicle; it's a lifestyle. And part of that lifestyle involves embracing the dirt, learning to appreciate the protective qualities of being perpetually covered in a layer of the great outdoors.

So, the next time you look at your mud-splattered, dust-coated Cybertruck, don't see it as a cleaning project. See it as a canvas, a living diary of your adventures. Each splatter of mud is a memory, each layer of dust a story. Embrace the dirt, for it is not just a sign of where you've been; it's a signal to the world of where you're going.

In essence, the dirty Cybertruck is the true Cybertruck. It's rugged, it's real, it's raw. It's a rolling testament to the road less traveled, the path less taken. So, go ahead, get it dirty. And then get it dirtier. Because in the world of the Cybertruck, the dirt isn't just dirt – it's a layer of experience, a coat of adventure, and a badge of honor. It's not something to be washed away; it's something to be cherished, celebrated, and above all, embraced.

# 10. Dubious Upgrades and Custom Jobs

Diving into the world of Tesla Cybertruck upgrades and custom jobs is akin to entering a bazaar full of shiny objects, where every stallholder assures you their wares are essential. In reality, many of these so-called upgrades are about as useful as a chocolate fireguard. The Cybertruck, with its design aesthetic that suggests it was dreamed up during a particularly intense sci-fi movie marathon, already looks like a prop from a budget-straining blockbuster. So, the idea of adding more bells and whistles to this already conspicuous vehicle is a path that can lead to both delight and despair.

Let's start with the most common of these upgrades – the off-road package. This is for those who believe that their Cybertruck, a vehicle that looks like it could bully a tank, needs to be more intimidating. This package usually includes tires so large they could have their own gravitational pull and a suspension lift that makes the truck so tall, getting into it without a ladder requires the athleticism of an Olympic gymnast. The result is a vehicle so imposing, so utterly over the top, that driving it makes you feel like you're leading a parade every time you leave the house.

Then there's the custom paint job. The Cybertruck comes in its natural stainless steel finish, a choice that suggests Tesla ran out of paint or simply forgot to order any. However, enterprising owners have taken it upon themselves to add a splash of color. The results range from tasteful and understated to designs so loud and garish they could be seen from space. It's like dressing a stealth bomber in a neon jumpsuit – sure, it's unique, but it's also a bit like screaming "look at me!" in a library.

Custom interiors are another avenue of personal expression. The standard Cybertruck interior, with its minimalist approach, has the charm of a high-end Scandinavian prison cell. To remedy this, some owners add custom upholstery and trim that make the interior look like the inside of a billionaire's private jet or, more often, like a teenager's bedroom with too much allowance. From diamond-stitched leather to wood trim so shiny it could be used as a mirror, these custom interiors can transform the truck from a utilitarian vehicle into something that screams, "I have more money than sense!"

Let's not forget the audio upgrades. The standard Cybertruck stereo is decent, but for some, 'decent' is not enough. They want a sound system that can double as a seismic event. These upgrades turn the truck into a rolling concert hall, capable of vibrating the fillings out of your teeth and ensuring that you and everyone within a half-mile radius can enjoy your music.

The pièce de résistance of dubious upgrades, however, has to be the external light bars. Because nothing says "I'm preparing for the apocalypse" quite like strapping a row of lights to your truck that could illuminate an entire football stadium. These lights are so bright, so utterly blinding, that they could probably be used as an interrogation tool.

In essence, the world of Cybertruck upgrades and custom jobs is a playground for those who believe that more is always better. It's where practicality goes to die, replaced by a mantra of excess and extravagance. Each upgrade takes the truck further away from its original purpose and transforms it into a rolling testament to the owner's personality – for better or worse.

So, if you're considering upgrading your Cybertruck, remember: just because you can, doesn't mean you should. But then again, in the world of the Cybertruck, sensible has never been part of the equation. It's a vehicle that invites excess, laughs in the face of subtlety, and gives a whole new meaning to the phrase "standing out from the crowd." Go forth, upgrade, customize, and turn your Cybertruck into a beacon of your unique taste. Just don't forget, at the end of the day, it still needs to fit in your garage.

## Adding a Spoiler for Absolutely No Reason

In the realm of the Tesla Cybertruck, where the laws of common sense and aerodynamics are flouted with the same disdain as a speed limit sign on the autobahn, the idea of adding a spoiler is not just unnecessary, it's gloriously absurd. It's like putting a saddle on a cow and expecting it to win the Grand National. But, in the ever-escalating world of vehicle customization, where impracticality is worn like a badge of honor, slapping a spoiler on the Cybertruck has become the automotive equivalent of wearing a top hat – it serves no real purpose, but it certainly makes a statement.

Now, let's consider the Cybertruck's design. It's as aerodynamic as a brick with aspirations. Its angular, sharp lines seem to have been inspired by a child's drawing of a car rather than anything resembling fluid dynamics. Adding a spoiler to this is like strapping a jet engine to a garden shed and expecting it to take off. The Cybertruck's relationship with aerodynamics is already tenuous at best; adding a spoiler is like trying to teach a whale to tap dance – an entertaining notion but ultimately futile.

But, let's say you decide to proceed with this endeavor, to boldly stick where no man has stuck before. You choose a spoiler – perhaps something subtle, or more likely, something that looks like it was ripped off the back of a car in a Fast & Furious movie. You attach it to the Cybertruck, standing back to admire your handiwork. The effect is immediate and dramatic. The truck now looks like it's trying to impersonate a sports car, like a hippo in a tutu attempting Swan Lake.

The beauty of this utterly pointless addition is not lost on the onlookers. As you drive down the street, heads turn, eyebrows raise, and a universal thought bubbles up: "Why?" But that's the point. In the world of the Cybertruck, 'why' is a redundant question. You didn't add the spoiler for any notion of improved performance; you added it for the sheer, unadulterated joy of excess, for the spectacle, for the theater of the absurd.

Driving with your new appendage, you notice no change in the handling, no improvement in fuel efficiency, and certainly no increase in speed. If anything, it's now harder to park, harder to fit in garages, and you're catching the wind like a sailboat. But none of that matters. What matters is the statement you're making, which is, essentially, "I can."

The spoiler, in its ostentatious glory, is a testament to human creativity, a homage to our endless pursuit of individuality. It's a talking point, a conversation starter, a visual pun. It's a reminder that in the world of car enthusiasts, sometimes logic takes a back seat to style, practicality bows down to personality, and common sense is left in the dust.

In essence, adding a spoiler to the Cybertruck is an exercise in whimsy, a foray into the realm of automotive theater. It's a nod to the absurd, a wink at the impractical, and a salute to the unnecessary. It's about taking a vehicle that already turns heads and adding your own personal twist, your own flair, your own nonsensical touch. So, embrace the ridiculous, celebrate the superfluous, and remember – in the grand pageant of life, sometimes it's not about need, it's about want. And if you want a spoiler on your Cybertruck, then by all means, add one. Just remember, it's about as useful as an ashtray on a motorcycle, but infinitely more entertaining.

## The Illusion of Speed: Convincing yourself that a spoiler makes your car go faster.

In the often illogical and ever-amusing world of the Tesla Cybertruck, adhering to conventional wisdom is about as common as a vegetarian shark. Enter the concept of adding a spoiler to this already unconventional vehicle – an endeavor akin to strapping a rocket to a tortoise in the hope of winning a sprint. The Cybertruck, with its geometric design that defies the very essence of aerodynamics, is as likely to benefit from a spoiler as a submarine is to benefit from a sunroof.

Yet, here we are, in a world where Cybertruck owners, armed with more optimism than a lottery player, attach spoilers to their electric leviathans. The rationale behind this is less about actual science and more about the delightful delusion that a piece of extraneous equipment can somehow magically transform their lumbering behemoth into a sprightly gazelle.

Let's break down the illusion. The Cybertruck, a vehicle that looks like it was designed using a ruler and a protractor, is not exactly a paragon of aerodynamic efficiency. Its shape, reminiscent of a stealth bomber if it were designed by a five-year-old, is about as conducive to air flow as a brick wall. Adding a spoiler to this mix is like putting a racing stripe on a bulldozer and expecting it to go faster.

But the power of belief is a strong force. You attach the spoiler, perhaps something sleek and aggressive-looking, an appendage that screams speed even when stationary. And then you take it for a spin. Lo and behold, despite the Cybertruck still adhering to the laws of physics, you're convinced it feels faster. The trees blur past with a touch more vigor, the wind howls with a tad more urgency, and you, the pilot of this electric titan, feel imbued with a newfound sense of velocity.

Of course, the actual effect of the spoiler on performance is about as significant as a homeopathic remedy is to medicine – it's all in the mind. The spoiler doesn't reduce drag; it doesn't increase downforce; frankly, it doesn't do much besides add a bit of weight and make fitting into parking spaces a game of Tetris. But in the driver's seat, fueled by a potent cocktail of placebo and pride, these facts are mere annoyances, easily swatted away like a fly at a barbecue.

This manual, with its usual blend of technical jargon and optimistic fluff, might make vague references to 'aerodynamic enhancements' and 'performance improvements.' These are, of course, terms to be taken with a pinch of salt – or perhaps the entire salt shaker. This manual knows, as do you deep down in the realms of reason, that the spoiler is as much a performance enhancer as a pair of high-top sneakers is to a marathon runner.

But in the grand theater of car ownership, where appearance often trumps reality, the spoiler plays a starring role. It's a statement, a conversation piece, a vehicular peacock feather. It says to the world, "I am here, I am bold, and I don't care for your conventional notions of speed and practicality."

So, as you cruise down the road, spoiler affixed and head held high, bask in the glory of your enhanced Cybertruck. Feel the rush of imagined speed, the thrill of perceived performance. The Cybertruck, after all, is not just a vehicle; it's a canvas for your automotive fantasies, a steel and electric embodiment of your desire to stand out, to defy norms, and to occasionally thumb your nose at the boring and the mundane.

In essence, the illusion of speed is just that – an illusion. But in the realm of the Cybertruck, where reality often takes a back seat to bravado, it's an illusion that brings joy, excitement, and a touch of rebellious flair. So let the skeptics scoff and the purists tut-tut; in your electric chariot, complete with its utterly unnecessary but utterly magnificent spoiler, you are the master of your own automotive destiny, a pioneer on the electric frontier, chasing the horizon with a smile on your face and a spoiler in your wake.

## Style Over Substance

When it comes to outfitting a Tesla Cybertruck with a spoiler, you're venturing into a realm where style triumphs over substance, where form forgets to follow function, and where practicality is left whimpering in the rearview mirror. Choosing a spoiler for your Cybertruck is less about enhancing performance and more about making a statement, akin to wearing a top hat at a beach party – it serves no real purpose, but it certainly gets you noticed.

Let's be clear: slapping a spoiler on the back of a Cybertruck is about as useful as installing a fireplace in a sauna. The Cybertruck, with its design that seems to have been inspired by a child's drawing of a tank, is as aerodynamic as a garden shed. Therefore, any spoiler you choose is not going to slice through the air with any more efficiency than a butter knife through concrete. But that's not why you're here. You're here because you want that spoiler to say something about you and your electric behemoth.

First, consider the size of your spoiler. This is crucial. Too small, and it's like whispering in a hurricane – utterly pointless. Too big, and you risk turning your Cybertruck into something that resembles a science fiction movie prop. You want a size that says, "Yes, I know this doesn't make my truck go faster, but it does make it look like it might."

Next, think about the style. Do you go for sleek and modern, a spoiler that looks like it was designed by an artist with a penchant for sharp angles? Or do you opt for something more aggressive, a spoiler that wouldn't look out of place on a race car, complete with fins and ridges that serve no purpose other than to look as if they might? Remember, the style of your spoiler is a reflection of your personality – choose wisely, or you risk sending the wrong message, like wearing socks with sandals.

Material is another consideration. Carbon fiber is a popular choice, lightweight and sporty, but let's be honest, on a Cybertruck it's like putting a silk hat on a pig. Alternatively, aluminum offers a more industrial, robust aesthetic, matching the Cybertruck's utilitarian vibe. It says, "I'm practical, but I also like to have pointless fun."

Color is where you can really express yourself. Do you match it to the Cybertruck's stainless steel hue, going for a look of understated (but pointless) elegance? Or do you choose a contrasting color, turning your spoiler into a visual shout that can't be ignored? Bold colors make a statement, but like a comedian at a funeral, make sure it's the right statement for the occasion.

Installation is the final hurdle. This is where you roll up your sleeves, get out the toolkit, and prepare to embark on a journey of frustration and skinned knuckles. This manual will suggest this is a straightforward process, but remember, this is the same manual that suggested the Cybertruck could withstand a sledgehammer. Be prepared for a battle, one where the only winners are stubbornness and perseverance.

In essence, choosing a spoiler for your Cybertruck is an exercise in futility, wrapped in a cloak of vanity. It won't make your vehicle faster, handle better, or improve fuel efficiency. But it will add character, flair, and a touch of personal style to a vehicle that already looks like it drove off the set of a post-apocalyptic blockbuster. It's about enjoying the aesthetics, embracing the absurdity, and acknowledging that sometimes, style really does triumph over substance. So go ahead, choose that spoiler, attach it to your Cybertruck, and drive with pride, secure in the knowledge that while it may not be useful, it is, without a doubt, moderately interesting.

## Customizing the Horn to Be Even More Annoying

In the thrilling and often eccentric world of Tesla Cybertruck ownership, the horn is not just a tool for warning the oblivious pedestrian or the dawdling driver. Oh no, it's a canvas for personal expression, a symphony waiting to be composed, a siren song that can oscillate between mildly irritating and ear-splittingly obnoxious. In the hands of a Cybertruck owner, the horn can be customized to reach new heights of annoyance, turning a simple beep into a cacophony that can shatter the peace of any tranquil neighborhood.

Firstly, let's address the factory-standard horn. In its original form, it's a polite little beep, a gentle reminder that you exist in the world. But where's the fun in that? As a Cybertruck owner, you don't do 'gentle' or 'polite.' You do grand, bold, and slightly absurd. Therefore, the first step in your quest is to discard the notion of the horn as a mere safety feature and instead see it as your personal loudhailer to the world.

Now, consider the options for customization. The most straightforward route is volume. By simply amplifying the decibels, your horn can go from a meek 'excuse me' to a thunderous 'move out of the way, peasant!' This is particularly effective in traffic jams, crowded parking lots, or when you want to announce your arrival at the local supermarket.

For the more melodically inclined, there's the option to change the tone of the horn. Why settle for a monotone beep when you can have a horn that plays the latest pop hit, a classic movie theme, or – for those with a penchant for the dramatic – the entire score of Wagner's 'Ride of the Valkyries'? Imagine the joy of pressing your horn and watching as everyone around you is treated to a full-blown orchestral performance emanating from your truck.

Then there's the option for a multi-tone horn. Why have one annoying sound when you can have several? With a multi-tone horn, each press can unleash a different sound. One moment it's a foghorn, the next it's a clown car, followed by a quack of a duck. It's like having a playlist for your horn, each track more infuriating than the last.

Of course, the pièce de résistance is the programmable horn. This is for the Cybertruck owner who wants complete control over their auditory arsenal. With a programmable horn, you can upload your own sounds. The cry of a howler monkey, the screech of a banshee, the wail of a siren – the possibilities are as endless as they are potentially headache-inducing.

Installation of these customized horns is, according to This manual, a straightforward affair. This, of course, is written in the same manual that suggests the Cybertruck can survive the apocalypse. Be prepared for a journey into the depths of the vehicle's electronics, a maze of wires and circuits that would baffle even Nikola Tesla himself.

In summary, customizing the horn of your Cybertruck is not just an upgrade; it's a statement. It says, "I am here, and you will hear me." It transforms the horn from a simple alert mechanism into an instrument of audio expression, a weapon of mass irritation. So go forth, choose your horn customization, and unleash it upon the world. Just be warned – with great volume comes great responsibility (and potentially a few noise complaints). But then, in the world of the Cybertruck, subtlety was never the point, was it?

## The Symphony of Irritation

In the bizarre world of the Tesla Cybertruck, picking the right horn sound is less about warning other road users and more about ensuring every living creature within a five-mile radius is acutely aware of your presence. The Cybertruck, with its imposing size and apocalyptic styling, already makes a statement. But why stop at visual intimidation when you can add an auditory assault to the mix? Choosing a horn for the Cybertruck is like picking the right ringtone for your phone, except instead of annoying just the person sitting next to you, you have the potential to irritate an entire postcode.

Firstly, let's consider the classic horn. Too mundane for the Cybertruck. The standard 'beep beep' is about as effective as a mouse squeak in a lion's den. No, Cybertruck owners need something with a bit more, shall we say, panache. Something that says, "I have arrived, and you will all now pay attention." Enter the world of custom horn sounds – a realm where subtlety goes to die.

Now, the logical choice might be something robust and truck-like – perhaps the deep bellow of a big rig, something that rumbles through the streets and vibrates in your chest like a bass guitar at a rock concert. It's assertive without being too flashy, a nod to the Cybertruck's utilitarian roots. But who are we kidding? This is the Cybertruck – utility is to this vehicle what salad is to a steak dinner.

So, you escalate. Maybe something theatrical? The blare of a trumpet, the wail of a saxophone, or for the more classically inclined, a snippet of Beethoven's Fifth Symphony. Each press of the horn unleashes a mini-concert, a musical interlude in the mundanity of traffic. It's not just a horn; it's an expression of your cultural flair. However, it's also a one-way ticket to being the most unpopular person in any given traffic jam.

But why stop at music? This is the Cybertruck, after all. You want a horn sound that truly stands out. Perhaps the roar of a lion, the scream of a jet engine, or – for those with a sense of humor – the sound of a clown car. Imagine the looks of confusion and annoyance as you roll through town with a horn that honks with the subtlety of a circus act.

The pièce de résistance, however, is the programmable horn. This is where you can truly let your imagination run wild. Record your own voice, shouting something suitably attention-grabbing. Or, for the ultimate in irritation, the sound of nails on a chalkboard. Nothing says 'notice me' quite like the collective cringe of an entire neighborhood.

Of course, with great power comes great responsibility (and the potential for a hefty fine for noise pollution). This manual, ever the proponent of mild anarchy disguised as innovation, encourages creativity but conveniently omits the potential for creating public enemies.

In conclusion, picking a horn sound for your Cybertruck is an exercise in auditory exhibitionism. It's about more than just making noise; it's about making a statement. Whether it's the roar of a wild animal, a blast of orchestral majesty, or the comedic beep of a clown car, your choice of horn sound is the aural embodiment of your personality. So choose wisely, or at least with a sense of mischief. After all, in the world of the Cybertruck, every journey should be an event, and with the right horn, you can be sure that your presence won't just be seen, but will be heard, felt, and – depending on your choice – either loved or loathed.

## Horn Etiquette, or Lack Thereof

In the wonderful world of the Tesla Cybertruck, where subtlety is as rare as a coherent tweet from a celebrity, the horn is not just a tool for safety – it's a weapon of mass distraction. Horn etiquette in a regular vehicle might involve a polite toot to alert a daydreaming pedestrian. But in the Cybertruck, equipped with a horn loud enough to wake the dead and startle the living, such niceties are as out of place as a vegan at a barbecue.

Firstly, let's establish the primary rule of Cybertruck horn etiquette: there is no etiquette. In the Cybertruck, the horn is there to be heard, much like a toddler with a new drum kit. The idea is not to gently nudge the attention of those around you; it's to grab their attention, shake it violently, and scream, "Look at me!"

Now, for the 'when'. The most obvious scenario is the classic city traffic jam. Here, amidst the cacophony of the urban jungle, your horn can shine – or rather, blare. It's not enough to simply press the horn; you must lean on it, giving it the full weight of your frustration. The goal is to create a sound so jarring, so utterly invasive, that it penetrates the very soul of every other driver, pedestrian, and unsuspecting bird in the vicinity.

Another prime opportunity is when encountering the meandering, oblivious pedestrian, the sort who views crossing the road as a leisurely stroll in the park. A quick blast of your horn should suffice to send them scurrying for safety, their leisurely stroll turned into a panicked sprint. It's like playing a real-life game of human pinball, with your horn serving as the flipper.

But why limit yourself to the mundane reality of traffic and pedestrians? The Cybertruck's horn is equally effective in more personal scenarios. Perhaps a friend hasn't noticed your arrival at their house. A courteous horn blast, loud enough to set off car alarms and rattle windows, is a surefire way to announce your presence. It says, "I'm here, and now it's a party."

Now, the 'how'. This is where finesse (or the Cybertruck version of it) comes in. The horn is not just an on-off switch; it's an instrument. A short, sharp blast says, "Excuse me, but I believe the light has turned green." A longer, more sustained note says, "Move, or I shall be forced to release the hounds." And for those truly special moments, a rhythmic pattern of honks can convey a whole range of emotions, from jovial greeting to apocalyptic rage.

Of course, This manual, in its typical tongue-in-cheek fashion, suggests using the horn "sparingly" and "with consideration for others." This is, of course, hilarious. In the Cybertruck, the horn is there to be used and abused, a mobile proclamation of your existence, a declaration of your presence in a world full of noise.

In essence, horn etiquette in the Cybertruck is an oxymoron. It's about as real as the tooth fairy or the concept of a quiet family dinner. The horn is there to be heard, to be feared, and to be revered. It's not just a part of the vehicle; it's an extension of your personality – loud, unapologetic, and impossible to ignore. So go forth, Cybertruckers, and use your horn with the same gusto with which the Cybertruck was designed. Just be prepared for the glares, the stares, and the occasional gesture that, while not polite, is certainly an acknowledgment that yes, you and your horn have been noticed.

## The Ill-advised Nitrous Oxide Button

In the absurdly entertaining world of the Tesla Cybertruck, there exists an option so preposterous, it could only have been conceived during a late-night, caffeine-fueled brainstorming session: the Nitrous Oxide button. This button, an ill-advised addition to any electric vehicle, let alone one shaped like a doorstop, promises to transform your sedate drive into a scene straight out of a low-budget action movie.

First, let's understand what we're dealing with here. Nitrous Oxide, commonly known as NOS, in the realm of sensible automotive engineering, is about as necessary as a chocolate teapot. In the Cybertruck, it's like giving a snail a jetpack – utterly unnecessary and likely to end in tears, but undeniably entertaining.

The Nitrous Oxide button in the Cybertruck sits there, a seductive red siren calling out to the driver. It whispers sweet nothings of unfathomable speed and power, promising to propel the hulking mass of stainless steel and battery cells into the realm of hypercars. Pressing this button is less an act of driving and more a leap of faith, a Hail Mary, a 'hold my beer' moment.

Now, the moment you press that button, the world changes. There's a sudden, eerie silence as the electric motors and batteries brace themselves. Then, with a whoosh that could awaken the dead, the Cybertruck lurches forward. It's as if you've just punched a hole in the space-time continuum, the scenery blurring into streaks of color like a child's finger painting gone rogue.

But, let's be clear – this is not a wise move. Engaging Nitrous Oxide in a vehicle already notorious for its hefty weight and questionable handling is akin to strapping an outboard motor to the Queen Mary and expecting it to perform like a speedboat. The Cybertruck, already as nimble as a dozing elephant, becomes a runaway freight train, careening down the road with all the grace of a tumbling boulder.

And then there's the aftermath. Once the euphoria of breakneck acceleration fades, reality sets in. The range of your battery plummets faster than the stock market on a bad day. The tires, designed for durability rather than drag racing, wear down with alarming speed. And the brakes, suddenly tasked with halting a runaway metal behemoth, age a decade in a matter of seconds.

This manual, with its usual blend of optimism and understatement, gently warns against 'frequent use' of the Nitrous Oxide system, noting that it may lead to 'reduced vehicle longevity.' This is akin to noting that jumping off a cliff may lead to a sudden stop. It casually neglects to mention that using NOS in a Cybertruck is about as sensible as using a flamethrower to light a candle.

In essence, the Nitrous Oxide button in the Cybertruck is a temptation, a test of your self-restraint, a red button begging to be pushed, but with consequences as predictable as rain in England. It's there not because it needs to be, but because in the world of the Cybertruck, excess is a virtue, and sanity is an afterthought.

So, should you press it? In moments of weakness, you might think, 'Why not?' But remember, with great power comes great responsibility – and potentially a great repair bill. The Nitrous Oxide button is a folly, a flight of fancy, a nod to the absurd. It's the automotive equivalent of a circus act – thrilling, breathtaking, and a little bit dangerous. Use it wisely, or better yet, just admire it, and leave the space-time continuum intact.

## More Fizz Than Fury

In the peculiar and often laughable universe of the Tesla Cybertruck, there exists a feature that's akin to a placebo in a pill bottle: the Nitrous Oxide, or NOS, button. This little red button, nestled among the various controls of the Cybertruck, promises the thrill of breakneck speed, the fantasy of turning a lumbering electric beast into a street-racing legend. However, the harsh reality is that this button is more about theatrics than actual mechanics, more fizz than fury, more akin to a magician's sleight of hand than a true performance enhancer.

Let's dissect this button, a small beacon of hope for those who dream of ludicrous speed. Pressing it conjures images of blasting down the road, leaving in your wake a blur of envious onlookers and a trail of bewildered pedestrians. But, much like expecting a political promise to materialize, pressing this button leads to a symphony of disappointment. The immediate realization dawns that you've been had – the only thing hurtling towards the horizon is your expectations.

The truth behind this mystical button is that it's essentially an automotive equivalent of a novelty whoopee cushion. It's there for a laugh, a bit of fun, a conversation starter over a pint. The Cybertruck, already about as aerodynamic as a garden shed, was never really destined to be a speed demon. The addition of a NOS button is like giving a sumo wrestler ballet shoes and expecting a pirouette.

When you do press the button, what happens? A slight whir, a subtle vibration, maybe even a gentle lurch forward, as if the Cybertruck is making a half-hearted attempt at humor. It's the vehicular equivalent of someone puffing out their chest and saying, "Hold my beer" only to open the beer and calmly sip it. The anticipated surge of acceleration is more imagined than experienced, a psychological jab rather than a physical uppercut.

And let's consider the implications of actually having a working NOS button in a vehicle like the Cybertruck. It's akin to strapping a jet engine onto the back of a tortoise. The Cybertruck, with its hefty battery pack and imposing stature, is about as suited to nitrous oxide-assisted speed as a hippo is to a high-wire act. The very thought of this button propelling the vehicle to warp speed is as laughable as the idea of a chocolate teapot.

This manual, with its usual blend of technical jargon and optimistic marketing speak, dances around the truth of the NOS button. It hints at increased performance, at exhilarating speeds, but these are the empty promises of a carnival barker. In reality, the NOS button is the automotive equivalent of those fake exhaust pipes you see on some cars – all bark and no bite.

In essence, the NOS button on the Cybertruck is a bit of fun, a whimsical nod to the fantasies of speed junkies and automotive dreamers. It's there to add a bit of spice to the otherwise utilitarian nature of the truck, a little bit of fizz in an otherwise serious concoction.

So, the next time you're behind the wheel of your Cybertruck and your gaze falls upon that tempting red button, remember what it represents. It's not a gateway to adrenaline-pumping speed; it's a nod to the child in all of us, the one that wants to believe in magic. Press it by all means, enjoy the momentary thrill, the theater of the mind that it invokes. But don't be disappointed when the laws of physics remain unbroken, and your mighty electric steed continues its steady, unflustered journey forwards. After all, in the grand pantomime of the automotive world, sometimes it's the illusion, the promise of what could be, that brings the most joy. And in the case of the Cybertruck's NOS button, it's definitely more about the fizz of excitement than the fury of speed.

## The False Rocket Launch

In the realm of the Tesla Cybertruck, where the lines between science fiction and reality are as blurred as the vision of someone walking out of a pub at midnight, there lies a feature so tantalizing yet so utterly disappointing – the so-called 'rocket launch' button. This button, which might as well be labeled 'Disappointment Guaranteed', is Tesla's equivalent of a practical joke, a faux feature that promises the excitement of a space launch but delivers the excitement of a space lecture.

Let's paint the picture. You're sitting in your Cybertruck, the vehicle that looks like it was designed by a child who was asked to draw a car but only had a ruler to hand. You see the button. It's there, shining with the promise of untold power, a beacon of hope in a world filled with mundane push-to-starts. You press it, your heart pounding with anticipation, your body ready for G-forces that could turn your face into a Picasso painting.

And then... nothing. Well, not nothing. There's a light. It lights up. That's it. It's the equivalent of expecting a fireworks display and getting a sparkler. It's like opening a door marked 'Treasure' and finding a note that says 'IOU one treasure'. The rocket launch button doesn't launch anything. It doesn't even launch a conversation; it just sits there, glowing smugly, as if to say, "Gotcha!"

This is the automotive equivalent of clickbait. It promises so much yet delivers so little. You half expect to press it and hear a voice saying, "This was a triumph. I'm making a note here: HUGE SUCCESS." But no, there's not even a sarcastic AI to keep you company. Just a light. A light that says, "You thought this would do something? How quaint."

The disappointment is palpable. It's like telling a child that you're taking them to Disneyland and then pulling up at a dentist. There's a sinking feeling, a sense of betrayal. You start questioning other things. Is the car even real? Are you real? Is anything?

Now, let's consider why this button exists. Perhaps it's a metaphor, a deep, philosophical statement about the unfulfilled promises of life. Or maybe, just maybe, it's Elon Musk having a laugh. It's him saying, "Let's see how many people I can trick into thinking their car can fly." If that's the case, well played, sir. Well played.

This manual, in a display of understated humor, refers to the button as an 'enhancement to the driving experience'. This is a bit like calling a whoopee cushion an 'enhancement to sitting down'. Sure, it adds something to the experience, but it's not quite what you were hoping for.

In essence, the rocket launch button in the Cybertruck is a lesson in managing expectations. It's a reminder that not all that glitters is gold. Sometimes, it's just a shiny button that lights up. It teaches us to find joy in the little things, even if that joy is accompanied by a side of disappointment.

So, the next time you press that button and nothing happens, just smile and remember: you're driving a vehicle that looks like it could survive the apocalypse. You don't need a rocket launch button to be special. Your car already looks like it was beamed down from an alien mothership. And in a world filled with boring sedans and uninspired hatchbacks, isn't that enough?

But if it's any consolation, remember this: while the button may not launch rockets, it does launch one thing – conversations. "What does this button do?" your passengers will ask. "Watch and find out," you'll reply, pressing it with all the ceremony of a NASA launch. And as the light comes on and their faces fall, you can bask in the glory of the false rocket launch, a feature that, in its own disappointing way, is truly out of this world.

# 11. Technical Mysteries

In the bizarre world of the Tesla Cybertruck, the term 'technical mysteries' doesn't quite cover the gamut of perplexities this vehicle presents. It's like calling the Bermuda Triangle a 'slightly tricky area.' The Cybertruck, with its design seemingly inspired by a child's drawing of a spaceship, is riddled with features and quirks that would have even the sharpest minds at NASA scratching their heads.

Let's start with the exterior. The Cybertruck looks like it was designed using a set square. It's all straight lines and sharp angles, a geometry teacher's dream. One look at it and you're compelled to wonder if it was Tesla's attempt to troll the entire automotive industry. "Let's see how many people we can convince to drive a stainless steel origami," Elon probably chuckled to himself. And lo and behold, the streets are now teeming with these angular monstrosities.

Moving inside, the mysteries deepen. The dashboard, for instance, is a minimalist slab of marble-like material, with a steering wheel that looks like it was nicked from a Formula 1 car. It's as if someone said, "Make it look futuristic, but also like it's been carved out of a single block of stone." The result is a dashboard that has less functionality than a chocolate teapot but looks like it belongs in a modern art gallery.

Then there's the central touchscreen. It's vast, like an iPad on steroids. This screen is your portal to understanding the Cybertruck, but it's about as user-friendly as a Rubik's Cube. It controls everything from the air conditioning to the radio, presumably even the trajectory if you decide to launch it into space. And navigating through its menus is like trying to solve the Da Vinci Code – you might eventually figure it out, but not before questioning your sanity.

The rear-view mirror is another technical conundrum. In keeping with the Cybertruck's futuristic theme, it's digital. This sounds impressive until you realize that it's about as clear as a fogged-up bathroom mirror. And adjusting it? You'd have an easier time trying to tune a radio in a Faraday cage. The result is a rear-view mirror that offers a view so pixelated, you're never quite sure if you're looking at a car behind you or an 8-bit character from a vintage video game.

And let's not forget the Cybertruck's electric motor and battery pack. With enough power to light up a small village, it's both impressive and baffling. The range is supposed to be impressive, but it fluctuates more than a politician's promises. One minute you've got enough range to cross continents, the next, you're nervously eyeing the battery indicator as it drops faster than your hopes and dreams.

In essence, owning a Cybertruck is like being part of an exclusive club whose initiation ritual involves deciphering an enigma wrapped in a riddle. It's a vehicle that prompts more questions than it answers. How does it work? Why does it look like that? What were they thinking? These are questions that Cybertruck owners whisper to themselves as they navigate the perplexing interface, adjust the cryptic mirrors, and drive their polygonal puzzle through the streets.

So, as a Cybertruck owner, embrace the mystery, revel in the confusion, and take pride in the fact that you're driving a vehicle that is part automobile, part spaceship, and part unsolvable puzzle. After all, in a world filled with mundane cars that make sense, isn't it a bit of fun to drive something that's a rolling technical mystery?

## Interpreting Warning Lights as Modern Art

In the bewildering world of the Tesla Cybertruck, interpreting the warning lights on the dashboard is less about vehicle maintenance and more about delving into a piece of abstract modern art. Each light, rather than providing clear and useful information, seems to have been designed to test your ability to interpret the most cryptic of hieroglyphics. It's as if Picasso and Dali had a love child, and that child decided to design car warning symbols.

Let's start with the classic 'check engine' light, a staple in any vehicle. In regular cars, this light indicates something is amiss under the hood. But in the Cybertruck, it's more enigmatic. This light could mean anything from "You forgot to close the frunk" to "I've decided to become a greenhouse and am currently cultivating a small rainforest in the boot." Trying to decipher its true meaning is like trying to understand the plot of a Christopher Nolan film on first viewing – you know it's profound, but you're not quite sure why.

Then there's the battery indicator light. In a standard EV, this would simply tell you that your battery is running low. But in the Cybertruck, this light flickers and changes colors with the randomness of a disco strobe light. It's less of a warning and more of an avant-garde light show. Is your battery low, or is the truck just trying to recreate the Northern Lights on your dashboard?

And let's not forget the tire pressure warning light. In the Cybertruck, this isn't just a simple indicator – it's a guessing game. The light comes on, but which tire is it referring to? Is it the front left, rear right, or perhaps the spare tire you forgot you had? It's like a game of Russian roulette, but with tires.

The most baffling of all, however, is the mystery icon that appears sporadically. It's just a series of squiggles and shapes that resemble an alien script. Is it a warning? A message from another galaxy? Tesla's way of keeping you on your toes? Deciphering this icon could be Elon Musk's way of encouraging driver alertness – you're certainly awake and alert when trying to figure out if your car is about to transform into a robot.

And when several of these lights come on at once, it's not just a warning – it's an art installation. Each light adds to the composition, creating a dazzling array of colors and patterns that could easily be displayed in the Tate Modern. Who needs a simple warning system when you can have an interactive light display that would make the Blackpool Illuminations look like a couple of tea lights?

In essence, the warning lights on the Cybertruck's dashboard are a masterclass in abstract art. They challenge your perceptions, provoke intrigue, and leave you with more questions than answers. They turn every drive into an interpretive dance, where you and the vehicle engage in a complex ballet of guesswork and intuition.

So, the next time those lights start flashing and flickering, don't panic. Instead, take a moment to appreciate the artistic genius before you. What does it mean? That's for you to interpret. After all, in the world of the Cybertruck, who needs straightforward when you can have enigmatic? It's not just a vehicle; it's a rolling modern art exhibit, complete with its own interactive light show. Just be sure to occasionally consult This manual – it may not be as illuminating as the lights themselves, but it might just stop you from mistaking a critical engine warning for an invitation to an art gallery opening.

## The Dashboard Disco

The dashboard of the Tesla Cybertruck, a vehicle that seems to have been designed by someone who thought the future would be a cross between a Mad Max film and a geometry textbook, is a bewildering array of flashing lights. These lights, rather than serving as helpful indicators of the vehicle's health, are more akin to a discotheque light show or a particularly abstract piece of modern art.

The Cybertruck's dashboard, when you first lay eyes on it, might lead you to believe that you've accidentally stumbled into an underground rave rather than a vehicle. Lights flash, colors change, and if you squint a bit, you might even see a strobe or two. It's as if the designers at Tesla, in a moment of madness, decided that what drivers really needed was not clarity or utility, but an in-car light show to rival the best nightclubs in Ibiza.

Let's start with the 'engine warning' light, a classic feature in most cars. In the Cybertruck, however, this light doesn't just gently inform you of a potential problem; it throws a full-on panic party. It flashes, it changes color, it might even pulsate to the beat of your increasing heart rate. It's less of a warning and more of an overzealous attempt to make you feel like you're driving a vehicle that's about to go into warp speed.

Then there's the battery indicator. In any sensible vehicle, this would simply tell you your charge level. But this is the Cybertruck, where sensibility is as rare as a concise Elon Musk tweet. The battery indicator doesn't just display your charge level; it puts on a performance. It fluctuates wildly, turning every trip into a nail-biting episode of "Will I Make It, or Will I End Up Stranded on the Side of the Road?"

And who could forget the tire pressure warning? In the Cybertruck, this warning light is more cryptic than the riddles of the Sphinx. It could mean anything from "You might want to check your tires" to "I've decided that I identify as a hovercraft now." The light doesn't just come on; it embarks on a rhythmic dance, leaving you to wonder whether you should check your tire pressure or start dancing along.

The pièce de résistance, however, is when all the lights decide to have a party. This usually happens when the Cybertruck is feeling particularly playful or when it senses that your life is devoid of excitement. Suddenly, every light on the dashboard flashes in unison, creating a kaleidoscope of color that could easily be mistaken for a piece from a modern art gallery. It's dazzling, it's confusing, and it's utterly useless as an indicator of anything, other than perhaps the car's aspirations to be a nightclub on wheels.

Interpreting these lights requires a shift in perspective. You must look not with the eyes of a concerned driver, but with the eyes of an art critic interpreting a particularly challenging piece. What does the flashing red light signify? Is it a commentary on the ephemeral nature of existence, or is it simply telling you that your seatbelt isn't fastened? The beauty, and the frustration, lies in the ambiguity.

In essence, the dashboard of the Cybertruck is less a functional display of important vehicular information and more an experimental art installation. It challenges your perceptions, tests your patience, and leaves you in a state of bewildered awe. So next time you're driving your Cybertruck and the dashboard lights up like a Christmas tree at a rave, take a moment to appreciate the spectacle. Just remember that beneath the dazzling display, there might just be an actual warning. Or not. In the Cybertruck, who really knows?

## Warning Lights or Christmas Decorations

When it comes to the Tesla Cybertruck, discerning whether the constellation of lights on the dashboard is a cause for concern or simply a festive display is like trying to understand why children eat glue: it's confusing, a little disturbing, and you're not quite sure what to do about it. In any normal vehicle, warning lights are just that – warnings. But in the Cybertruck, they're more akin to a set of Christmas lights, randomly blinking in a way that's both mesmerizing and mildly concerning.

Let's start with the light that looks like a nuclear reactor. In a regular car, this might suggest a catastrophic engine failure, the kind that makes you think fondly of your bank balance before it's annihilated. In the Cybertruck, however, it could mean anything from "You're low on washer fluid" to "I fancy a game of Pictionary." It's less of a warning and more of a conversation starter.

Then there's the mysterious triangle with an exclamation mark – the automotive equivalent of a shrug. In normal cars, this light generally means something is amiss, but it's not immediately clear what. In the Cybertruck, it's Tesla's way of keeping you on your toes. It's like playing a game of automotive charades. Is it the brakes? The electrical system? Or has the Cybertruck just become sentient and is now pondering the meaning of its existence?

And let's not forget the tire pressure warning light, which in the Cybertruck flickers with the whimsy of a candle in a breeze. In most vehicles, this light is a straightforward indication to check your tires. In the Cybertruck, it's more of a gentle suggestion, like a butler discreetly whispering in your ear that you might want to consider the possibility of looking at your tires at some point in the not-too-distant future, if it's not too much trouble.

But the real pièce de résistance is when all the lights decide to join the party. This usually happens at the least opportune moment, like when you're navigating rush hour traffic or showing off your new electric beast to a group of awestruck friends. Suddenly, the dashboard transforms into a rave, a symphony of blinking, flashing lights, each one vying for your attention. It's like driving inside a Christmas tree, if the Christmas tree was having a bit of an existential crisis.

Now, the Cybertruck manual, in all its optimistic ambiguity, suggests that these lights are part of the vehicle's charm, a feature, not a bug. This is akin to saying that a sudden downpour at a picnic is an 'unexpected water feature.' Sure, it adds to the experience, but not necessarily in a good way.

The challenge, then, is deciding whether to be concerned about these lights or to simply sit back and enjoy the show. It's a fine line between vigilance and blissful ignorance. On one hand, you don't want to be the person who ignored the warning lights and ended up stranded on the side of the road with a Cybertruck that's decided to take an unscheduled nap. On the other hand, you don't want to be the person who runs to the mechanic every time the dashboard decides to throw a disco party.

In conclusion, navigating the world of Cybertruck warning lights is about balance. It's about understanding that while some of these lights might be important, others are just Tesla's way of adding a bit of spice to your daily drive. It's about not sweating the small stuff, but also about knowing when the small stuff might just be a harbinger of a larger, more expensive problem. So, the next time your Cybertruck's dashboard lights up like a festival, take a moment to appreciate the spectacle, but also spare a thought for what those lights might be trying to tell you. After all, in the world of the Cybertruck, every day is Christmas, but sometimes, Santa brings a lump of coal.

## The Random Reset: Automotive Roulette

In the perplexing and often bizarre world of the Tesla Cybertruck, there exists a phenomenon as mystifying as it is maddening – the random reset. This is not your garden-variety glitch or your run-of-the-mill hiccup. No, this is the Cybertruck's version of playing automotive roulette, a game where the stakes are as high as the vehicle's inexplicably angular design.

Picture the scene: you're cruising along, the envy of every other road user, basking in the glory of your electrically powered polygon. The sun is shining, the birds are singing, and all is right with the world. Then, without warning, everything goes dark. The dashboard lights extinguish, the touchscreen goes blank, and you're left in a state of confusion so profound, it's akin to finding out your vegan burger was made of real beef all along.

This, dear Cybertruck owner, is the random reset, a feature so unpredictable, it makes British weather look consistent. It's like playing a game of Russian roulette with your vehicle's electronics, except instead of a bullet, it's your sanity that's at stake.

The first time it happens, you might pass it off as a fluke, a one-off anomaly. But as it recurs, with the regularity of a politician's scandal, you come to realize that this is just part of the Cybertruck experience. It's Tesla's way of adding a bit of excitement to your day, like a jack-in-the-box that's lost its spring. You never know when it's going to pop up and startle you into next week.

But let's delve into the mechanics of this random reset. One moment, everything's working as smoothly as a ballroom dancer on ice. The next, it's as if someone's pulled the plug on your Cybertruck, leaving you in a state of high-tech limbo. It's a complete system reboot, a Ctrl-Alt-Del on a vehicular scale. And when it comes back to life, it's like waking a sleeping giant – slow, groggy, and slightly disoriented.

Navigating the random reset requires a combination of patience, a sense of humor, and a strong resistance to the urge to bang your head against the nearest solid object. It's a test of your character, a trial by fire, or in this case, a trial by electronics.

The cause of these resets? Who knows. It could be anything from a software glitch to an overly sensitive sensor, or perhaps the Cybertruck is just sentient and enjoys a good laugh at your expense. This manual, in its typical fashion, glosses over these resets with the vagueness of a politician avoiding a direct question. It suggests 'turning the vehicle off and on again', the automotive equivalent of giving it a good shake and hoping for the best.

In essence, the random reset in the Cybertruck is a reminder that technology, no matter how advanced, still has the capacity to throw a tantrum like a toddler denied candy. It keeps you on your toes, adds a bit of unpredictability to your journeys, and serves as a conversation starter, albeit one that begins with, "You won't believe what happened to me on the way here..."

So, the next time your Cybertruck decides to play automotive roulette and resets itself, just remember: it's all part of the adventure. It's the price you pay for driving a vehicle that looks like it was designed for a Mars mission. Embrace the chaos, enjoy the unpredictability, and always keep your sense of humor about you. After all, in the world of the Cybertruck, anything can happen – and frequently does.

## Discovering the joys of your car's systems resetting at the most inopportune moments.

In the world of the Tesla Cybertruck, the term 'surprise reboot' is not so much a technical term as it is a euphemism for "Oh no, not again." It's like having a butler who randomly decides to vanish just as he's about to pour your tea. This feature, if one can call it that, adds a thrilling element of unpredictability to your driving experience, much like playing a game of musical chairs, except the music is your engine, and the chair is your entire vehicle.

Imagine you're cruising along the motorway, the pride of owning such an angular monstrosity swelling in your chest. The dashboard looks like something out of Star Trek, the touchscreen is gleaming, and you're starting to feel like Captain Kirk on wheels. Then, suddenly, the screen blinks out, the dashboard goes dark, and you're left wondering whether you've accidentally stumbled into a wormhole.

This, dear driver, is the surprise reboot – the Cybertruck's way of reminding you that, despite its futuristic facade, it's still just a car, and cars, as we know, have a mind of their own. The first time it happens, you might think it's a fluke. The second time, a coincidence. But by the third, fourth, and fifth times, you'll start to wonder if your Cybertruck is possessed.

The beauty of the surprise reboot lies in its impeccable timing. It never happens when you're safely parked and admiring your reflection in the shop window. No, it prefers to strike at the most inopportune moments – like when you're navigating a particularly tricky roundabout or in the middle of overtaking a tractor on a country lane. It's as if the car has a mischievous spirit that enjoys a bit of drama.

Now, the reasons for these reboots could range from the mundane to the mystifying. Maybe it's a software glitch, or perhaps the car's computer is overwhelmed by the sheer absurdity of its existence. Or maybe, just maybe, Elon Musk is watching from a secret control room, pushing buttons and cackling to himself. We may never know.

But it's not just the reboot itself that's the issue; it's the aftermath. Once the car comes back to life, it's like waking up a grumpy bear from hibernation. The systems light up one by one, each with its own series of beeps, boops, and bongs, as if it's trying to communicate in Morse code. The touchscreen might come back with all the speed of a lazy Sunday morning, leaving you to wonder whether it's time to start planning your route via the stars.

And let's not forget This manual's take on this. In a masterpiece of understatement, This manual suggests 'waiting patiently' for the systems to reboot. This is akin to telling someone who's just fallen overboard to 'wait patiently' for the tide to bring them back.

In essence, the surprise reboot in the Tesla Cybertruck is one of its most endearing quirks, a reminder that in an age of ever-advancing technology, we are still at the mercy of the humble computer reboot. It adds a frisson of excitement to every journey, a story to tell at every dinner party.

So, the next time your Cybertruck decides to take a brief nap mid-journey, just remember: it's all part of the experience, a tale to tell, a memory to cherish. After all, in a world where everything is predictable, isn't it nice to have a bit of surprise? Just maybe not while you're on the motorway, please.

## Reset and Hope for the Best

In the eccentric world of the Tesla Cybertruck, the 'reset and hope for the best' feature, or as some might call it, the automotive equivalent of Russian roulette, is as thrilling as finding a wasp in your underpants. There's nothing quite like the adrenaline rush of not knowing whether your vehicle will remember who you are, where you're going, or if it even believes itself to be a car anymore.

Imagine you're driving along, the Cybertruck purring like a cat that's swallowed a synthesizer, and suddenly, the dashboard lights up like a Christmas tree in Times Square. The infotainment system decides to take a brief siesta, and you're left in a state of suspense that would make Alfred Hitchcock proud. This is the Cybertruck's random reset – a feature that tests your patience, your sanity, and your ability to remember what all your settings were.

After what seems like an eternity (but is probably only a few minutes), the system springs back to life with the enthusiasm of a Monday morning commuter. Now comes the moment of truth – what has survived the great digital apocalypse? Will your satellite navigation remember your destination, or will it now believe you're navigating the streets of Mars? Will your carefully curated playlist still be there, or has it been replaced by the sound of Elon Musk laughing?

The random reset doesn't just reboot the system; it rolls the dice on your settings. It's like a game show, but instead of winning a prize, you're playing not to lose your sanity. Each time is a mystery. Maybe your seat will remember your preferred settings, or perhaps it'll now position itself in a way that would make a chiropractor wince. Maybe your mirror will still be angled just right, or maybe it'll now be more useful for checking if there's any wildlife sneaking up on you from the backseat.

And then there's the climate control. Before the reset, it was set to a comfortable 21 degrees Celsius. Now, it's either recreating the conditions of the Sahara or the Arctic – you're either melting into your seat or considering whether it's appropriate to wear a ski suit for the drive.

This manual, in its infinite wisdom, offers sage advice like "resetting may cause temporary loss of functionality." This is a bit like saying, "Stepping outside in a thunderstorm may cause temporary wetness." It's not just a loss of functionality; it's a loss of faith in all the technological advances we've supposedly made since we discovered fire.

In essence, the random reset on the Cybertruck is not just a feature; it's an adventure. Each time it happens, you're not just resetting the car; you're resetting your expectations of what driving a car should be like. It's a reminder that, in this age of technology, sometimes the machines have a mind of their own.

So, the next time your Cybertruck decides to take a brief journey into the abyss of a reset, just remember: it's all part of the experience. Embrace the uncertainty, the suspense, the sheer thrill of not knowing what comes next. After all, life is full of surprises – and so is your Cybertruck. Just maybe keep a map and a compass handy, just in case your electric steed decides it's a 19th-century carriage post-reset.

## When to Simply Give Up and Call a Tow Truck

In the maddening world of the Tesla Cybertruck, there comes a point when the only sensible course of action is to throw in the towel, wave the white flag, and call a tow truck. It's like admitting defeat at a pub quiz when the next category announced is 'Quantum Physics.' The Cybertruck, with its more quirks than a family reunion, sometimes leaves you no choice but to admit defeat in the face of its baffling eccentricities.

The first sign that it's time to make that humiliating call is when the Cybertruck decides to play dead. You're sitting there, key in hand, a hopeful glint in your eye, and then nothing. Not even a whimper. It's as if the Cybertruck has decided to enter witness protection and didn't tell you. You prod, you poke, you even utter a few choice words, but it's as responsive as a teenager after a heavy night out.

Then there's the moment when the dashboard resembles the control panel of the Starship Enterprise during a Klingon attack. Lights are flashing, alarms are sounding, and you're half expecting Scotty to beam you up. This manual, which up until now has been as much use as a chocolate fireguard, tells you to 'consult a professional.' This is manufacturer speak for "Even we don't know what's going on, mate."

Another surefire sign is when the Cybertruck's electric motor develops a personality. One minute, it's purring along, smoother than a jazz singer in a velvet suit. The next, it's making a sound like a cat being strangled. This isn't just an engine problem; it's an exorcism waiting to happen. At this point, not only should you call a tow truck, but perhaps also a priest.

And let's not forget the pièce de résistance: the total system shutdown. This is the automotive equivalent of the blue screen of death on your computer. Everything goes dark, the touchscreen is as blank as a politician's promises, and you're left with a vehicle as mobile as a concrete bollard. The Cybertruck has turned into an expensive driveway ornament, and no amount of button pushing or colorful language will bring it back to life.

Of course, there's always the hope that it might just be a flat battery – a simple, understandable problem. But in the back of your mind, you know it's something more. It's the Cybertruck's way of saying, "I've had enough. It's not me, it's you."

So, when do you give up and make the call? When you've tried everything in This manual, everything your mechanically inclined friend suggested, and everything you found on a Tesla forum at 3 a.m. When your toolbox is empty, your patience is spent, and your neighbors have started to ask if you're starting a collection of large, angular paperweights.

In essence, knowing when to call a tow truck for your Cybertruck is like knowing when to leave a party. If you stay, you might fix the problem, but you're more likely to end up tired, frustrated, and questioning your life choices. So, swallow your pride, pick up the phone, and call for that tow. It's a humbling experience, but remember – even the most advanced electric vehicle is still a machine. And machines, much like humans, sometimes just need a little help.

So, there you have it. Your Cybertruck, a marvel of modern engineering, a beacon of the electric future, sometimes just needs a good old-fashioned tow. It's not a defeat; it's a tactical retreat. A chance to regroup, reassess, and return to fight another day. And who knows? Maybe the tow truck driver will have some interesting stories to tell.

## The Inevitable Surrender

In the grand, perplexing saga of the Tesla Cybertruck, there comes a time when you, the intrepid owner, must face the music, or in this case, the blinking lights of doom on your dashboard. It's the moment when you realize that, despite your best efforts, your beloved Cybertruck has thrown a spanner in the works so complex, so unfathomable, that it makes solving a Rubik's cube blindfolded look like child's play.

Let's set the scene: You're out on the open road, the wind is in your hair, or it would be if the Cybertruck had any discernible aerodynamic properties. The polygonal beast is humming along, then suddenly it isn't. The dashboard lights up like a Christmas tree in a power surge, the touchscreen display starts flickering with all the stability of a British government, and the motor makes a sound that suggests it's attempting to communicate in Morse code.

Your first instinct, armed as you are with a YouTube degree in car mechanics, is to fiddle with a few things. You poke around under the hood, which in the Cybertruck is about as useful as looking under your sofa cushions for the meaning of life. You consult This manual, a tome filled with as much helpful information as a cookbook in a famine. After a few hours, or days (who's counting?), you realize that you're about as likely to fix this problem as you are to teach your cat quantum physics.

That's when it hits you: The Inevitable Surrender. The realization that sometimes, just sometimes, you need to call in the professionals – those brave souls who make a living deciphering the hieroglyphics of automotive engineering. It's a humbling moment, one that gnaws at the very heart of your can-do spirit. But it's either that or turn your Cybertruck into the world's most expensive lawn ornament.

So, you make the call. The mechanic arrives, and you watch, a knot of anxiety and hope in your stomach, as they delve into the guts of your electric beast. They poke around, mutter a few technical terms that sound like they're straight out of Star Trek, and occasionally shake their head in a way that does nothing for your confidence.

As you stand there, you start to reflect on the journey you've taken with your Cybertruck. From the moment you laid eyes on it, with its design that looks like it was inspired by a child's drawing of a car after watching too much sci-fi, to the first time you got behind the wheel, feeling like the king of the electric road. It's been a relationship filled with highs, lows, and a smattering of bewildering moments that defy logic.

And in this moment of reflection, you come to appreciate the value of professional help. Sure, there's a certain romance to fixing things yourself, the satisfaction of solving a problem with your own two hands. But there's also a certain charm in not spending your weekends with your head buried in an engine bay, your hands covered in grease, and your soul filled with existential dread.

In essence, accepting the need for professional help with your Cybertruck is not admitting defeat. It's admitting that you have better things to do with your time, like actually driving your Cybertruck, rather than trying to divine why it's suddenly decided to impersonate a very large and expensive paperweight.

So, embrace The Inevitable Surrender. See it as an opportunity to learn, to grow, and to hand over your problems to someone who might actually know what they're doing. And remember, in the world of cutting-edge electric vehicles, sometimes the cutting edge cuts back. When that happens, it's okay to step back and let the professionals take over. After all, they're the ones with the toolbox that doesn't just contain a hammer and a hopeful expression.

## Tow Truck Tales

Owning a Tesla Cybertruck is akin to being in a tumultuous relationship with a high-maintenance partner. It's all glitz and glamour until it leaves you stranded on the hard shoulder, wondering where it all went wrong. The Cybertruck, with its futuristic design that looks like it was sketched by a hyperactive child with a ruler, is as much a marvel of modern technology as it is a masterclass in unpredictability.

Let's start with the classic 'out of charge' scenario. You'd think with all the dials, displays, and whatnot, keeping an eye on the battery life would be as easy as spotting a toupee in a strong wind. But no, the Cybertruck likes to keep you guessing. One moment, you're sailing along with enough range to circumnavigate the globe, or so it seems. The next, you're gliding to an unceremonious stop, the battery as empty as a politician's promises.

Then there's the time the Cybertruck decides to play an elaborate game of dead. You get in, ready for a leisurely drive. You press the start button, but the truck remains as lifeless as a doorknob. You try everything – sweet-talking it, cursing it, pleading with it. But it's no use. The Cybertruck has checked out, and you're left with no choice but to call the tow truck.

Ah, the tow truck – the Cybertruck owner's reluctant acquaintance. You've seen them more times than you've had hot dinners. They arrive with the same look of bemusement as someone who's just seen a dog walk on its hind legs. "It's the future, eh?" they chuckle, as they hook up your motionless pride and joy.

But it's not just the stopping and starting that leaves you stranded. Oh no, the Cybertruck is more creative than that. There was that one time when the autopilot got a bit too enthusiastic and took you on an impromptu tour of the countryside. Or the time when the doors decided they preferred being art installations, refusing to open and turning the truck into an impenetrable fortress.

Let's not forget the infamous software update debacle. You're driving along, and a message pops up suggesting an update. "Why not?" you think. It's just like updating your phone, right? Wrong. Halfway through the update, the Cybertruck decides it's had enough, leaving you stranded in the technological equivalent of no man's land, with a truck as functional as a chocolate teapot.

Each stranded episode ends the same way – you, standing on the roadside, waiting for the tow truck, the Cybertruck sulking behind you like a grounded teenager. You've started to recognize the tow truck driver; maybe even added them on social media. "See you next week!" they call out as they drive away, and you can't help but chuckle. It's either that or cry.

In conclusion, owning a Cybertruck is not just about enjoying the cutting edge of electric vehicle technology. It's about the stories, the adventures, the "you'll never believe what happened to me" tales. It's about standing on the side of the road, waiting for the tow truck, and knowing that, despite it all, you wouldn't trade your Cybertruck for the world. Because let's face it, who wants boring when you can have a vehicle that keeps you – and the local tow truck company – on your toes?

# 12. Warranty Weaseling

Delving into the warranty of the Tesla Cybertruck is akin to reading a mystery novel where the last page, which reveals whodunnit, is conveniently missing. You see, in the fantastical realm of the Cybertruck, where the vehicle looks like a rejected prop from a low-budget sci-fi movie, the warranty is a complex labyrinth designed to be as clear as mud on a rainy day.

First, let's consider the basic premise of a warranty – a promise that if things go south, you're covered. Simple, right? Wrong. The Cybertruck's warranty is so riddled with clauses, conditions, and caveats that it would make a tax attorney weep. It's like playing a game of Snakes and Ladders, except every ladder is broken, and the snakes are on steroids.

Now, you might think, "Ah, but my Cybertruck is a robust beast, what could possibly go wrong?" Oh, how naive. Remember, this is a vehicle that once had its windows smashed in a demonstration of their un-smash-ability. The point is, things will go wrong. But whether those things are covered by the warranty is a question that could baffle even the Oracle at Delphi.

Let's say, for instance, your Cybertruck's door handle stops working. You check the warranty. Door handles? Covered. But wait! Upon closer inspection, and after consulting a magnifying glass and the Rosetta Stone, you discover that door handles are only covered if they stop working on a Tuesday, under a full moon, and only if you've never played a Justin Bieber song in the vehicle.

And then there's the battery – the heart of the Cybertruck. Surely, the warranty on the battery is straightforward? Ah, but you forget, this is the Cybertruck. The warranty on the battery is more conditional than a celebrity's apology. The battery is covered, but only if you can prove that you've never charged it on days ending in 'y', never driven uphill, and always recited a haiku before starting the vehicle.

But the pièce de résistance of the warranty weaseling is the paint job. The Cybertruck comes in any color you want, as long as it's stainless steel. So, you'd think, "No paint, no paint warranty issues," right? Wrong again. There's a clause in there about maintaining the structural integrity of the stainless steel. It's covered, but only if you regularly polish it with the tears of a unicorn and never expose it to direct sunlight, indirect sunlight, or the gaze of a jealous neighbor.

Of course, let's not forget the fine print. Oh, the fine print. It's written in a font so small, you'd need an electron microscope to read it. It's here that you'll find all the things that aren't covered by the warranty, which, as it turns out, is pretty much everything. It's like buying a waterproof phone that's only waterproof if you never get it wet.

In essence, understanding the warranty of your Cybertruck is like trying to solve a Rubik's Cube blindfolded, with one arm tied behind your back, while riding a unicycle. It's a herculean task that's both mentally and emotionally draining. So, when something does go wrong with your Cybertruck, as it inevitably will, you have two choices: try to navigate the warranty's treacherous waters, or simply give up and start a new life in a remote village where the word 'warranty' is met with confused stares. After all, ignorance is bliss, especially when it comes to the Cybertruck warranty.

## Limited Warranty: Very Limited Indeed

In the fantastical world of the Tesla Cybertruck, the term 'limited warranty' takes on a new, almost whimsical meaning. It's the kind of warranty that's so limited, it makes a hermit look sociable. It's like receiving an invitation to an all-you-can-eat buffet, only to arrive and find nothing but a plate of stale crackers. This warranty is a masterclass in the art of giving with one hand while taking away with the other – and then running away with the hand.

Let's start with what the warranty ostensibly covers. It suggests a protective blanket over your beloved Cybertruck, a promise that, should anything go wrong, Tesla will be there with bells on, ready to fix it. But as you delve deeper, you realize that this blanket is more of a handkerchief, and a small one at that.

The Cybertruck's body – a geometric marvel that looks like it was designed by a ruler-wielding madman – is covered, but only against rust. Rust! In a vehicle made of stainless steel. That's like offering sunscreen to a fish. It's the automotive equivalent of offering a warranty against shark attacks in the Sahara.

Then there's the battery warranty. Oh, the battery – the heart and soul of the Cybertruck. The warranty here is about as reassuring as a chocolate teapot. Sure, it's covered, but only if you haven't charged it too much, or too little, or at the wrong times, or on days ending in 'y'. It's like saying, "You're covered, as long as you don't actually use your car."

But the pièce de résistance is the warranty's handling of the electrical system. This is where things get so convoluted, it makes a game of Twister look straightforward. The warranty covers the electrical system, but with so many exclusions and caveats, you wonder if it only applies when the planets are aligned. It's like being covered for all illnesses, except for every illness known to mankind.

And let's not forget the software – the Cybertruck's brain. Here, the warranty tiptoes around issues like a ballet dancer in a minefield. Software problems are covered, but only those that Tesla acknowledges as problems. This is the equivalent of a doctor saying, "Yes, you're sick, but we only cover illnesses that we find interesting."

Now, for the pièce de résistance: the 'act of God' clause. This little nugget absolves Tesla from any responsibility if your Cybertruck is damaged by something outside human control. This includes, but is not limited to, alien invasions, spontaneous combustion, or an attack by a flock of particularly aggressive pigeons.

What does the warranty actually cover then? Well, if you read between the lines, squint hard, and maybe have a few glasses of wine, you'll see that it covers the air in the tires, the dust on the dashboard, and any spiders that may have taken up residence in the glove compartment.

In essence, the limited warranty of the Cybertruck is a masterclass in optimistic marketing. It's like being promised a trip to the moon, only to be taken to a moon-themed party at the local pub. It gives with one hand, takes with the other, and then nicks your wallet while you're not looking.

So, when you buy a Cybertruck, remember that the warranty is very limited indeed. It's there, sort of, in the background, like a distant relative at a wedding. It might occasionally be useful, but most of the time, it's just there for decoration. And when things go wrong, as they inevitably will, just remember: you didn't just buy a car, you bought an adventure – and adventures rarely come with a warranty.

# Discovering the many, many exclusions of your "comprehensive" warranty.

When you dive into the 'comprehensive' warranty of the Tesla Cybertruck, you'll find it's about as comprehensive as a vegetarian menu at a steakhouse. This warranty, a document so riddled with exclusions it could be used as a colander, is a joy to behold – if your idea of joy is discovering you've bought a chocolate teapot.

First off, let's address the elephant in the room – or rather, the elephant not in the room, because it's likely excluded from the warranty. You see, the Cybertruck's warranty has more holes in it than the plot of a bad soap opera. It promises the world, but delivers an atlas – a nice picture, but not quite the real thing.

For starters, the warranty on the battery – the heart, soul, and spleen of the Cybertruck – is a marvel of legal wordplay. It's covered, sure, but only if you treat it like a delicate flower that's never seen the harsh light of day. Charge it too often or not enough, and you might as well have used it as a doorstop. The warranty essentially says, "We trust you to use the battery, but not really."

Then there's the paintwork, or in the case of the Cybertruck, the lack thereof. Any issues with the stainless steel finish? Covered, but only if you can prove you haven't exposed it to air, water, or the gaze of a jealous neighbour. It seems the only way to keep the warranty valid is to store your Cybertruck in a vacuum-sealed bubble, away from the dangers of the outside world, like a fragile piece of moon rock.

And let's talk about the electrical system. This warranty covers everything, as long as everything is defined as 'practically nothing'. If a bulb goes out, you're covered – but only if that bulb was never turned on in the first place. It's like saying you're covered for all illnesses, as long as you've never been ill.

But the pièce de résistance, the cherry on this cake of despair, is the drivetrain warranty. It covers all the moving parts – unless they move. Any sign that you've actually used the Cybertruck for, you know, driving, and the warranty waves goodbye faster than your sanity on a bad day.

Let's not forget the infamous 'Act of God' clause, a delightful little get-out-of-jail-free card for Tesla. If your Cybertruck is damaged by something beyond human control – say, a butterfly flapping its wings in Japan – then it's adios warranty. This clause is so broad; it could include alien abductions, time travel mishaps, or an invasion of sentient toasters.

The fine print of the Cybertruck's warranty is where hopes and dreams go to die. It's a labyrinth of legalese, a jungle of jargon. Wading through it is like trying to do a crossword puzzle in a language you don't speak, blindfolded, with both hands tied behind your back.

In essence, the 'comprehensive' warranty of the Cybertruck is as comprehensive as a book on the history of the universe written by a goldfish. It promises much but delivers little. It's a safety net with holes big enough to drive a Cybertruck through – sideways. So, when you buy your Cybertruck, read the fine print, have a laugh, and then put it away. After all, you didn't buy a Cybertruck for the warranty. You bought it for the thrill, the adventure, and the sheer joy of explaining to confused onlookers exactly what it is they're staring at. And let's face it, that's worth more than any warranty.

## Warranty or Lottery

In the grand, unpredictable circus that is owning a Tesla Cybertruck, understanding the warranty is akin to playing a game of chance. It's like buying a lottery ticket, except the prize is not a life-changing sum of money, but rather the slim chance that whatever just broke on your Cybertruck might actually be covered. The warranty, a document so riddled with loopholes it could double as a script for a legal drama, is about as clear as a mud bath.

First, let's consider the battery, the beating heart of the Cybertruck. The warranty suggests that it's covered, but reading the fine print reveals a different story. The battery is covered, provided you haven't charged it more than necessary, or too little, or on days when Mercury is in retrograde. Basically, if your battery fails, you'd have better luck getting blood from a stone than getting a claim approved.

Moving on to the Cybertruck's robust stainless steel body, which looks like it was designed by a geometry teacher with a vendetta against curves. You'd think this would be covered comprehensively. Oh, how naïve! The body is covered unless it encounters anything harder than a marshmallow. Did a pebble kicked up by a passing bicycle nick the body? That's on you. It seems the only way to keep the warranty intact is to suspend the Cybertruck in a vacuum chamber, untouched by human hands.

Now, let's talk about the tires - those round rubber things that occasionally touch the ground. You'd expect these to be covered, considering they're fairly crucial to the whole 'driving' experience. But the warranty views tires as a whimsical accessory, like a novelty air freshener. Pothole damage, wear from actually using your vehicle for its intended purpose, or even just looking at them funny, and the warranty evaporates faster than your patience on a Monday morning.

Electrical systems? Well, they're covered, but only in the event that they fail while the car is switched off, parked, and possibly in another dimension where electrical systems are impervious to failure. If your touchscreen develops a fault or your electric windows decide they prefer being walls, you might as well try fixing them with positive thoughts and pixie dust.

The paint job, or lack thereof, on your Cybertruck is a tricky one. Seeing as the vehicle looks like it's been carved out of a single block of metal by a vengeful robot, you'd think paint issues wouldn't exist. But oh, they do. If by some miracle you manage to scratch it, the warranty treats it as a personal affront and washes its hands of you completely.

Then there's the drivetrain. This crucial series of components is theoretically covered by the warranty, but in practice, it's about as reliable as a chocolate teapot. If anything goes wrong, the warranty's stance is that it must have been caused by driving the vehicle - which, last time you checked, was kind of the whole point.

The most fun part of the Cybertruck warranty, however, is the 'Act of God' clause. This little gem effectively exempts Tesla from any responsibility if your Cybertruck is damaged by something outside of human control. This includes, but is not limited to, asteroid strikes, dragon attacks, or an impromptu performance of Riverdance by a herd of wild deer on your roof.

In essence, the warranty of the Tesla Cybertruck is less a warranty and more a whimsical guessing game. It's like playing 'pin the tail on the donkey', except the donkey is a mirage, and the tail is made of unobtainium. So, as you navigate the perilous waters of Cybertruck ownership, remember: the warranty is not so much a safety net as it is a decorative piece of paper, designed to give you a false sense of security while you gamble on the reliability of one of the most eccentric vehicles ever to grace the road. Good luck – you'll need it.

# Service Schedules: More Like Guidelines

In the world of the Tesla Cybertruck, service schedules are treated with about as much reverence as a speed limit sign on the Autobahn. These schedules, ostensibly provided to keep your prized electric behemoth running like a dream, are less rigid timetables and more like those guidelines that pirates are so fond of. They're the automotive equivalent of a weather forecast in Britain – take it with a grain of salt and always prepare for the opposite.

Firstly, let's address the frequency of these service schedules. This manual suggests a series of regular check-ups, presumably to ensure that the Cybertruck hasn't spontaneously decided to morph into a toaster. However, the Cybertruck, much like a rebellious teenager, doesn't adhere to schedules. One day it's fine, the next it's displaying more warning lights than a Christmas tree in Times Square.

Then there's the matter of what these service schedules entail. Typical cars require oil changes, filter replacements, and various fluid top-ups. But the Cybertruck scoffs at such mundane necessities. Its idea of a service is akin to a spa day – a bit of software pampering here, a battery massage there, and perhaps a gentle cleansing of its sensors. It's less about keeping the car running and more about keeping it happy – because apparently, cars have feelings now.

However, attempting to stick to these service schedules is like trying to nail jelly to a wall. You bring your Cybertruck in for its check-up, and the technician, with a look of bewilderment, hooks it up to a computer that looks like it's capable of launching a space shuttle. The screen lights up with a list of diagnostics that makes the control panel of the Starship Enterprise look user-friendly.

Now, you'd think that with all this technology, the service would be thorough and precise. But no, it's more of a general once-over, a cursory glance, a 'yeah, it looks alright' kind of deal. It's as if the technician is just as confused by the Cybertruck as you are, but neither of you wants to admit it.

And let's talk about the costs of these services. You'd expect, with a vehicle as futuristic as the Cybertruck, that the servicing costs would be astronomical. But surprisingly, it's more like throwing a few coins into a wishing well. You're never quite sure what you're paying for, but you hope it brings good luck. The receipt lists items like 'firmware caress' and 'battery ego boost', which, while amusing, are about as enlightening as a black hole.

What's more, these service schedules seem to operate in a parallel universe where time is a mere concept. A 'quick check-up' can range from an hour to what feels like a decade. You sit in the waiting room, sipping on stale coffee, flicking through magazines from 2005, and wondering if you've somehow slipped into a temporal anomaly.

In conclusion, service schedules for the Cybertruck are more like guidelines – loose, flexible, and open to interpretation. They're a dance, a delicate ballet between what Tesla thinks your Cybertruck needs and what the Cybertruck decides it will tolerate on any given day. So, approach these service schedules with an open mind, a sense of humor, and perhaps a packed lunch – because you never know how long you'll be in for. And remember, in the world of the Cybertruck, it's not just a service; it's an adventure – a vague, confusing, sometimes expensive adventure, but an adventure nonetheless.

## Understanding that service schedules are more suggestions than actual rules.

When it comes to the Tesla Cybertruck, the service schedules are about as rigid as a politician's promise. These schedules, presented in a manual that appears to have been written by someone who thinks time is a relative concept, are less concrete plans and more like a horoscope reading – vaguely applicable to everyone and specific to no one.

The Cybertruck, a vehicle that looks like it was designed using a ruler by someone who failed art class, operates on its own understanding of time and maintenance. The idea that you could have a set schedule to service this metallic origami is as ludicrous as expecting a cat to adhere to a daily planner. It's an exercise in optimism, a testament to human hope over experience.

Now, let's delve into these service schedules. You have your typical checkpoints: 10,000 miles, 25,000 miles, and so on. But in the world of the Cybertruck, these numbers are mere suggestions, whimsical figures plucked from the ether. They assume a world where the Cybertruck behaves like a regular vehicle, adhering to the laws of automotive physics. But, as any Cybertruck owner will tell you with a haunted look in their eyes, this truck laughs in the face of convention.

The first service schedule arrives, and you dutifully bring in your Cybertruck, expecting a routine check. But the moment the technicians hook it up to their diagnostics, the truck decides to throw a curveball. Error codes that make as much sense as a drunken Morse code message flash up, sensors that were functioning perfectly now decide to have an existential crisis, and you're left wondering if your truck was possessed by the spirit of a disgruntled 80s computer.

Then there's the issue of Tesla's over-the-air updates, a futuristic concept that's about as reliable as a weather forecast during British summertime. These updates promise to keep your Cybertruck in tip-top condition, but they often leave it more confused than a grandparent using a smartphone. One day, your truck is purring like a kitten; the next, it's meowing like it's just seen a cucumber, convinced that its battery is now a toaster.

This manual, bless its optimistic soul, suggests that adhering to these schedules will ensure the longevity and happiness of your Cybertruck. This is akin to saying that eating apples will keep doctors away – a quaint idea, but not one grounded in the gritty reality of life with a Cybertruck. The truth is, these service schedules are more like a game of pin-the-tail-on-the-donkey, blindfolded, after a few pints.

So, what does this mean for the average Cybertruck owner? Essentially, it means embracing the art of approximation. Your service schedule might say 10,000 miles, but in Cybertruck language, that could mean anything from 8,000 to 12,000 miles, or whenever the truck decides it's had enough and starts blinking at you like a confused owl.

This laissez-faire approach to maintenance might seem alarming at first, but it's all part of the Cybertruck experience. It's about learning to listen to the truck, to understand its quirks and its moods. The Cybertruck doesn't just want a mechanic; it wants a confidante, a partner in crime, someone who understands that sometimes, it just wants to be left alone.

In conclusion, the Cybertruck service schedules should be taken with a pinch of salt, a shrug of the shoulders, and a healthy dose of humor. They are not so much strict timelines as they are a series of hopeful suggestions, a guideline written in the sand at low tide. And in this way, owning a Cybertruck is not just about driving a vehicle; it's about embarking on a journey of discovery, of learning to live with a machine that's as unpredictable as it is innovative. Just remember to keep your sense of humor about you – you're going to need it.

## Learning why putting off service can be a risky game.

Owning a Tesla Cybertruck and procrastinating on its service is like playing a game of automotive chicken. It's the vehicular equivalent of waiting until the fuel light is not just on, but flashing Morse code for "Help me." You know you should get it serviced, but there's always something more pressing, like watching paint dry or teaching your dog to recite Shakespeare.

The Cybertruck, with its design that looks like it was dreamed up during a particularly intense cheese dream, doesn't just need a service; it demands it, in the same way a diva demands only green M&Ms in her dressing room. But let's face it, when you have a vehicle that resembles a stealth bomber on wheels, it's easy to assume it's invincible. Spoiler alert: it isn't.

Procrastinating on servicing your Cybertruck is a gamble where the stakes are as high as Elon Musk's ambitions to colonize Mars. Sure, it might run smoothly for a while, lulling you into a false sense of security. But beneath its stainless-steel exterior, there's a labyrinth of software and mechanics that's just waiting to throw a tantrum. And when it does, it won't just be a simple hissy fit. No, it'll be a full-blown, operatic meltdown.

The first sign that your game of procrastination is about to come crashing down is usually something small. Maybe the auto-pilot starts getting a bit too literal about the 'auto' part and tries to steer you into a hedge. Or perhaps the touch screen starts acting like a stroppy teenager, refusing to respond to your touch unless you say please and thank you.

But things can escalate quickly. One minute you're cruising along, the wind in your hair, smug in the knowledge that you've dodged the servicing bullet. The next, your Cybertruck is making a noise like a Tyrannosaurus Rex with indigestion. It's then that you realize, as the dashboard lights up like a Christmas tree, that perhaps, just perhaps, you should have taken it in for that service.

And let's not forget the updates. Oh, the updates. The Cybertruck, with its constant need for software updates, is like a needy pet that constantly requires attention. Ignore these updates at your peril. Because if you do, one day you might find that your Cybertruck has decided to identify as a submarine and is trying to re-route your journey via the nearest large body of water.

Procrastination also leads to some interesting conversations with the mechanic. You sheepishly admit that you've been putting off the service, and they look at you as if you've just confessed to thinking the world is flat.

Then they give you the bill, and you realize that procrastination doesn't just cost you in stress, but also in cold, hard cash.

In essence, putting off servicing your Cybertruck is like playing a game of Jenga with your sanity. It's all fun and games until the whole thing comes crashing down. The Cybertruck may be a marvel of modern technology, but it's also a machine. And machines, much like humans, need a little TLC every now and then.

So, embrace the service schedule. Treat it like a spa day for your Cybertruck. It'll thank you for it, and you'll avoid those heart-stopping moments when you start to wonder if your car is about to spontaneously combust. After all, owning a Cybertruck should be about the joy of the drive, not the fear of what happens when you push your luck just a little too far.

## Customer "Support": An Exercise in Patience

In the realm of owning a Tesla Cybertruck, engaging with customer support is akin to entering a Kafkaesque realm where logic goes to die and patience is stretched thinner than a politician's excuse. It's an adventure, a test of wills, a foray into a world where the word "support" is as loosely interpreted as a salad at a Texas barbecue.

Imagine, if you will, your Cybertruck suddenly decides to express its artistic side and paints its touchscreen display with a dazzling array of error messages. You, being a reasonable person, think a quick call to customer support will resolve this. Ah, such sweet naivety.

First, you must navigate the labyrinthine maze that is the automated phone system. It's a series of cryptic choices, each one leading you further into the abyss of hold music that sounds like it was composed by a depressed robot. By the time you reach an actual human being, you've aged several years, and your hairline has begun a tactical retreat.

Then, there's the conversation with the customer support representative – a person who has mastered the art of sounding sympathetic while simultaneously giving the impression they'd rather be anywhere else. They listen to your plight, clack away on a keyboard for what seems like an eternity, and then offer the kind of advice that makes you wonder if they're actually trying to fix your problem or write the next great American novel.

"Have you tried turning it off and on again?" they ask, a question so clichéd it makes you wonder if you've accidentally called the set of a sitcom. You suppress a sigh, explain that yes, you have indeed tried the technological equivalent of a Hail Mary, and wait for their next pearl of wisdom.

What follows is a dance, a delicate ballet of back-and-forth, where you try to explain that your Cybertruck is having a meltdown, and they try to find new ways to tell you it's probably your fault. "Have you exposed the car to sunlight?" they ask, as if you're keeping a vampire in your garage. "Did you speak harshly to the touchscreen?" because obviously, the Cybertruck has the emotional resilience of a Victorian damsel.

Eventually, after what feels like a stint in purgatory, you're offered a solution so convoluted, it requires a PhD in Cybertruck-ology to understand. It involves rebooting, recalibrating, and possibly performing a ritual dance around the vehicle at midnight. You hang up, armed with a list of tasks that makes the labors of Hercules look like a walk in the park.

But it's not just the phone calls. Oh no. If you're particularly masochistic, you might venture into the online support forums, a place where hope goes to die. Here, you'll find a collection of souls, all wandering lost and confused, asking questions into the void and receiving answers that range from wildly unhelpful to borderline mystic.

In conclusion, dealing with customer support for your Cybertruck is not just a call; it's a journey, an epic saga. It's an exercise in patience, a test of your resolve, a quest for answers in a sea of uncertainty. But it's also a reminder that at the end of the day, you're part of an exclusive club. A club of pioneers, braving the new world of electric vehicles, armed only with your wits and a Cybertruck that's as unpredictable as it is revolutionary. So, take heart. Embrace the challenge. And always remember, the next time you call customer support, bring a snack – it's going to be a long one.

## Finding inner peace while listening to the same tune for hours on end.

In the grand adventure of owning a Tesla Cybertruck, few experiences test the limits of human endurance like the dreaded customer service hold music. It's the auditory equivalent of being stuck in a lift with an accordion player who only knows one song. This isn't just hold music; it's a descent into a special kind of purgatory where time loses meaning and sanity hangs by a thread.

Let's set the scene. You've encountered a problem with your Cybertruck, a vehicle that looks like it was designed by a geometrist after a particularly heavy night out. You pick up the phone and dial customer service, armed with hope and a naive belief in the efficiency of corporate helplines. That's your first mistake.

As soon as your call is answered by an automated voice, you know you're in for the long haul. The voice presents you with a menu of options, each one leading inexorably to the same destination: the hold queue. It's like choosing the form of your own torture. Do you want the quick and painless option that doesn't exist, or the slow descent into madness?

You choose an option, and that's when it starts. The music. Oh, the music. It's a tune that sounds like it was composed by someone who hates music, happiness, and probably puppies. It's a melody that loops every 30 seconds, a maddening earworm that burrows into your brain and lays eggs.

At first, you try to ignore it. You make a cup of tea. You scroll through your phone. But the music is relentless. It follows you, haunts you, becomes the soundtrack to your slow descent into madness. It's like being trapped in a lift with the world's most boring musician playing the world's most boring song.

As the minutes turn into hours, you start to wonder if this is some sort of psychological experiment. How long can a human being listen to the same tune before they crack? You start to imagine the hold music as a physical entity, a malevolent sprite dancing just out of sight, cackling at your misery.

But it's in these darkest moments that something miraculous happens. You begin to find a sort of zen-like peace in the endless loop of music. You transcend your earthly concerns, your anger, your frustration. You become one with the hold music. It's no longer a tune; it's a mantra, a meditative chant that guides you to a higher plane of existence.

In this new enlightened state, you realize that the hold music is not your enemy. It's a teacher, a guru, a guide on your path to inner peace. It's teaching you patience, endurance, and the ability to find serenity in the face of adversity. You're no longer just a Cybertruck owner; you're a warrior, a survivor, a veteran of the hold music wars.

When the customer service representative finally picks up the line, you're almost disappointed. The real world seems harsh, jarring, after the soft cocoon of the hold music. You solve your Cybertruck issue with the calm detachment of a monk, your soul now tempered by the fires of endless, maddening hold music.

In conclusion, the next time you're subjected to the endless hold music of the Tesla customer service line, don't despair. Embrace it. Let it guide you to a new level of understanding. For in the great cosmic joke that is owning a Cybertruck, the hold music is perhaps the greatest teacher of all.

## Scripted Responses and Frustration

Dealing with customer service for the Tesla Cybertruck is a bit like trying to have a conversation with a parrot that's been trained to recite only Shakespearean insults. You're not going to get very far, and you'll probably end up more confused than when you started. The customer service experience is a baffling maze of scripted responses, each more infuriatingly unhelpful than the last, delivered with the emotional range of a teaspoon.

When you first call up with a problem – let's say your Cybertruck's door has decided it prefers being an immovable wall – you're greeted with a level of enthusiasm usually reserved for someone who's just been told they're about to undergo a root canal. "Your call is very important to us," drones the voice on the other end, in a tone that suggests they're staring into the abyss and contemplating the futility of existence.

Then you're launched into the abyss of the script. It doesn't matter what your problem is – whether your Cybertruck has transformed into an impromptu BBQ grill or has started communicating in Morse code – the script remains unflinchingly, maddeningly the same. "Have you tried turning it off and on again?" they ask, as if you hadn't already considered this technological panacea.

But it's not just the banality of the questions that grinds your gears; it's the relentless cheerfulness with which they're delivered. Each suggestion is offered with the kind of false joviality that makes you wonder if the person on the other end isn't a robot powered by crushed dreams and a perpetual existential crisis.

Occasionally, you might try to break through the script with a joke or a bit of light-hearted banter, but it's like trying to make a brick wall laugh. The customer service rep sticks to their script with the determination of a limpet clinging to a rock in a storm. You could say, "I think my Cybertruck is haunted," and they'd respond with, "Have you checked the tire pressure?"

Navigating this maze of scripted responses requires the patience of a saint and the sense of humor of a stand-up comedian performing at a funeral. It's an exercise in futility, an endless loop of the same questions and answers, spiraling into a vortex of despair. You half expect them to start asking if your refrigerator is running or if you have Prince Albert in a can.

And just when you think you've reached the end of your tether, when you're about to unleash a tirade that would make a sailor blush, you remember: this is all part of the unique charm of owning a Cybertruck. It's not just a vehicle; it's a lifestyle choice, a commitment to a future where cars look like they've been designed by a five-year-old with a ruler and a love of sci-fi, and where customer service is an elaborate performance art piece.

So, you take a deep breath, plaster a smile on your face, and dive back into the script. "Yes, I've tried turning it off and on again. Yes, I've checked the tire pressure. No, I haven't tried singing it a lullaby and tucking it in at night, but I'll give that a go too." After all, when you own a Cybertruck, life is too short to be taken seriously. Especially when it comes to customer service.

# 13. Addendum of Afterthoughts

In the unlikely event that you've survived the main body of this 'Cybertruck Owner's Manual' without descending into madness or fleeing to a remote island, welcome to the Addendum of Afterthoughts. This is where we cram in all the bits and pieces that didn't quite fit anywhere else, much like the last items you shove into a suitcase before sitting on it to close. This section is the junk drawer of This manual – a collection of oddities and curiosities about your beloved Cybertruck that are as perplexing as they are pointless.

Firstly, let's discuss the Cybertruck's paint. Or rather, its lack thereof. Tesla, in a move of sheer audacity, decided that paint was too mainstream and went with a stainless steel finish that looks like it was borrowed from a rejected kitchen appliance. The lack of paint means you don't have to worry about color-matching after scratches. Instead, you can enjoy the dulcet tones of metal scraping against, well, everything, as you try to navigate a car park.

Then there's the issue of parking. With its angular design, the Cybertruck doesn't so much glide into a parking space as it does threaten it. Parking this beast is like trying to fit a square peg into a round hole, if the square peg was made of stainless steel and could glare menacingly at passersby. And don't even think about multi-story car parks – unless you fancy turning the roof into an accidental tribute to origami.

Let's not forget the towing capacity, a feature Tesla boasted about with all the subtlety of a bull in a china shop. While the Cybertruck can indeed tow an impressive amount, what they don't mention is that doing so will drain your battery faster than a bathtub with no plug. You might be able to pull a small moon, but only for about ten miles before you need a recharge.

The infotainment system deserves a special mention. It's like having an oversized iPad bolted to your dashboard, except with fewer functions and more confusion. Navigating through the menus requires the skill and precision of a brain surgeon, and the likelihood of accidentally playing 'Baby Shark' on full blast is disturbingly high.

As for the autopilot feature, it's a bit like having a well-meaning but slightly inebriated friend take the wheel. It works, sort of, but you can't shake off the feeling that it might drive you into a hedge at any moment for a laugh. It's recommended to keep your hands close to the steering wheel, and perhaps a foot hovering over the brake, just in case your Cybertruck decides it's a good time to explore the scenery up close.

Finally, we arrive at the pièce de résistance: the Cybertruck's ability to float. Yes, you read that right. According to Elon Musk, it can float. While this may sound like a marvelous feature, it's probably wise to remember that just because it can float, doesn't mean it should. The Cybertruck is many things, but a boat is not one of them. Unless you fancy a very expensive experiment in marine archaeology, keep it on dry land, where its bewildering design and perplexing features can be fully appreciated by confused onlookers.

In conclusion, the Cybertruck is not just a vehicle; it's a rolling testament to human ambition, creativity, and perhaps a touch of madness. It challenges the norms, pushes boundaries, and leaves a trail of bewildered and amused bystanders in its wake. Owning a Cybertruck is not just about getting from A to B; it's about the journey, the experience, and the stories you'll have to tell. Just remember to take everything in this manual, and indeed about the Cybertruck, with a healthy dose of skepticism, a pinch of salt, and a good sense of humor.

## List of Unusable Tools Included with Your Vehicle

In the world of the Tesla Cybertruck, the phrase 'tool kit' takes on a whole new meaning, and not in a good way. The tools included with your vehicle are as useful as a chocolate teapot, and about as durable. Let's embark on a whimsical journey through this collection of unusable tools, each more perplexing than the last, and all seemingly designed by someone who's only ever seen tools used in cartoons.

First up, we have the ludicrously tiny wrench. This little gem is so minuscule, it's only really suitable for tightening the screws on a child's toy, or perhaps for performing delicate surgery on a hamster. The idea that it could be used on the robust nuts and bolts of a Cybertruck is laughable. It's like bringing a toothpick to a sword fight.

Then there's the screwdriver, or as I like to call it, the 'screw-loosener'. It's made from metal so soft, you'd think it was crafted from recycled aluminium foil. The first time you apply any real pressure, it twists like a contortionist at a yoga retreat. It's about as effective as trying to dig a hole with a spoon.

Let's not forget the hammer – an object that bears a passing resemblance to a real hammer but performs with about as much effectiveness as a rubber chicken. It's the perfect tool for gently tapping in a nail, as long as the nail is made of butter and you're hammering it into a cake.

Next, we have the pliers – a tool that seems to have been designed by someone who's never actually used pliers. Their grip is weaker than a newborn kitten's, and the first time you try to grip anything tougher than a marshmallow, they give up faster than a lazy person's New Year's resolution. They're about as much use as a pair of chopsticks in a soup-eating contest.

And then there's the pièce de résistance: the 'universal' socket wrench. This tool promises the world and delivers none of it. It's supposed to fit any bolt, but in reality, it fits none. It's like a key that's been designed to open every door but ends up opening none. It's the Loch Ness Monster of the tool world: much talked about, but when you need it, nowhere to be found.

The inclusion of a tape measure would normally be cause for celebration, but not in this case. This tape measure is so flimsy, it recoils at the mere thought of measuring anything longer than a few inches. It's less a tape measure and more a limp piece of ribbon with delusions of grandeur.

Lastly, we have the flashlight, a tool that emits about as much light as a glow worm on a foggy night. It's perfect if you're trying to navigate the vast expanse of the Cybertruck's interior in pitch darkness and only want to see a fraction of an inch in front of you. It's the kind of flashlight you'd give to someone you didn't like if you wanted them to get hopelessly lost.

In essence, the tool kit provided with your Tesla Cybertruck is an exercise in optimism over experience. Each tool is a testament to hope over practicality, a promise of functionality that falls flat the moment it's put to use. So, when you first lay eyes on this collection of tools, remember to laugh. Because if you don't laugh, you might cry, and tears will get you nowhere when you're trying to fix a flat tire with a chocolate teapot of a wrench.

## A comprehensive guide to tools that are for display purposes only.

Welcome to the world of the Tesla Cybertruck, where the toolkit provided is about as useful as a chocolate fireguard or a knitted condom. This isn't so much a toolkit as it is a collection of metallic sculptures, a modern art exhibit that you can store in your garage. These tools are to DIY what I am to vegetarianism: fundamentally opposed on every conceivable level.

Let's start with the hammer – a glorious, shiny object that looks like it was forged in the fires of Mount Doom, but in reality, couldn't knock a nail into a block of butter. If you ever find yourself in a situation where hammering is required, I recommend using any part of the Cybertruck itself – it's bound to be more effective.

Then we have the screwdrivers, a diverse range of exquisitely crafted metal sticks with heads that strip faster than a burlesque dancer on fast forward. If you're planning on actually using these to screw or unscrew anything, you might as well be trying to perform brain surgery with a spork. Their real purpose is to be laid out on a workbench to impress your friends with your apparent preparedness for vehicular maintenance.

Now, onto the pliers. These beautifully polished instruments are perfect for gently holding a sugar cube for your morning tea. Their grip strength is comparable to that of an arthritic octogenarian. They're so ineffectual that you'd have more luck using telekinesis to hold a bolt in place.

The adjustable wrench in the toolkit is a marvel of engineering incompetence. It has the unique ability to adjust itself to precisely the wrong size, regardless of the bolt. It's like a shapeshifter with an identity crisis. If you actually manage to make this tool grip something, buy a lottery ticket immediately, because it's your lucky day.

Of course, no decorative toolkit would be complete without a set of Allen keys. These are provided in sizes that do not correspond to any known bolt on the Cybertruck. Their purpose is a mystery, one that not even Alan Turing could decipher. They are the Loch Ness Monster of the tool world – often talked about but never seen serving any practical purpose.

The pièce de résistance of this toolkit is the tape measure. It is a delicate, fragile thing, capable of retracting with such force that it could create a black hole. The measurements are in a unit that Tesla seems to have invented specifically for this purpose, rendering it as useful as measuring distance in spaghetti strands.

Finally, we have the flashlight, designed to emit a beam of light so feeble that it could be outshone by a glow-in-the-dark sticker. This is the tool you use when you want to navigate your way not out of, but further into, darkness. It's the equivalent of trying to signal for help with a candle in a hurricane.

In conclusion, the toolkit provided with your Tesla Cybertruck is not so much a toolkit, but a test – a test of your patience, your ingenuity, and your ability to keep a straight face while you pretend that these tools might actually be of some use. They are decorative, aspirational, a symbol of what could be if they weren't so utterly, irrevocably useless. So hang them on your garage wall, admire their aesthetic qualities, and then, when it's time to actually fix something, call a professional. Or better yet, use the Cybertruck itself – it's the toughest tool you'll own.

## Understanding why the supplied jack doesn't actually fit your car.

In the grand, often bewildering world of Tesla Cybertruck ownership, the supplied accessories are a constant source of amusement, confusion, and in some cases, outright bewilderment. Among these, the car jack stands out – or rather, fails to stand up – as a shining example of mismatched optimism. This isn't so much a tool as it is a test of faith, a puzzle that would leave even the Sphinx scratching its head.

The jack, a contraption that looks like it was designed by someone who once saw a car jack in a cartoon and thought, "That doesn't look too hard to make," is a marvel. It's as if Tesla took the concept of a car jack, threw it into a blender with some abstract art and a bit of wishful thinking, and then fashioned the result into something resembling a car accessory.

Firstly, the size. The Cybertruck, a vehicle so large it could double as a small apartment, comes with a jack that looks like it was meant for a toy car. Attempting to lift the Cybertruck with this jack is like trying to heave Moby Dick onto a fishing boat with a slingshot. It's an exercise in futility, optimism over reason, David versus Goliath if David had forgotten to bring his sling.

Then, there's the issue of compatibility. The jack fits the Cybertruck with about as much precision as a pair of wellington boots fits a snake. It's akin to buying trousers without trying them on, only to find they fit everywhere except where they should. Using this jack requires not only strength, determination, and a bit of engineering know-how, but also a healthy dose of blind hope and possibly a few fervent prayers.

Let's not forget the design of this jack, which is a masterpiece of impracticality. It's as if the designers were given the brief to create a car jack that defies not just expectations, but also the basic laws of physics. The jack seems to operate on the principle of wishful thinking rather than any established mechanical guidelines. It's a piece of equipment that promises so much yet delivers so little, like a politician's election campaign.

Moreover, the instructions for the jack are a saga in themselves. Written in what appears to be an ancient, forgotten dialect, with diagrams that resemble a toddler's first attempt at origami, they only add to the overall sense of confusion. One can't help but feel that successfully operating this jack would qualify you for a degree in mechanical engineering or perhaps even wizardry.

And then there's the experience of actually using the jack – a tale of adventure, danger, and unexpected acrobatics. Positioning the jack under the Cybertruck is a task that would make Hercules think twice. Once in place, the act of lifting the vehicle is akin to performing a ballet while wrestling a bear. It's a Herculean task that requires not just physical strength, but also a strong will to live.

In conclusion, the car jack supplied with the Tesla Cybertruck is less a tool and more a token, a nod to the idea that sometimes, it's the thought that counts. It's a testament to Tesla's unique approach to problem-solving: provide the tools, but make sure they're more symbolic than practical. So, when you find yourself in need of lifting your Cybertruck, remember to approach the task with a sense of humor, a dash of creativity, and perhaps a backup plan – like a nearby crane.

## Incomprehensible Software Licensing

Diving into the software licensing of the Tesla Cybertruck is like trying to untangle a bowl of spaghetti with a pair of chopsticks. It's a bewildering, headache-inducing foray into a world where logic goes to die and confusion reigns supreme. The licensing agreements are so dense, so convoluted, that attempting to understand them is like trying to read War and Peace, backwards, in the dark, while someone shouts random numbers at you.

Let's start with the basic premise: you own a Cybertruck, but do you own the software that makes it more than just a giant, angular paperweight? The short answer is no. The long answer is also no, but with about fifty pages of legal jargon that makes less sense than a baboon performing brain surgery. The software in your Cybertruck is licensed, not sold, which means you're essentially renting it from Tesla – like a high-tech landlord who can evict you at any time for reasons that are as mysterious as they are numerous.

The terms and conditions of this software license read like a dystopian novel. There are clauses, sub-clauses, and sub-sub-clauses, each one more confusing than the last. It's a labyrinth of legalese, a quagmire of qualifications. You need a law degree, a Rosetta Stone, and possibly a stiff drink to make any sense of it.

For instance, there's the delightful clause that states Tesla can update, change, or revoke the software at any time, for any reason. It's like waking up one morning to find your Cybertruck now identifies as a submarine and refuses to drive on land. And if you have a problem with that? Well, you can always try reading the software license agreement again and descend further into madness.

Then there's the small matter of what you can and can't do with the software. Modifying it is out of the question, unless you fancy turning your Cybertruck into a very expensive brick. Even attempting to peek at the software's inner workings is forbidden – it's like owning a mystery box that you're never allowed to open, under pain of being haunted by the ghost of Elon Musk.

But the real cherry on top of this legal sundae is the clause about data. By driving a Cybertruck, you agree to hand over just about every conceivable piece of data to Tesla. Where you go, how fast you drive, what you had for breakfast – okay, maybe not that last one, but give it time. It's like having a nosy neighbor living in your glove compartment, taking notes and occasionally sending them to Big Brother.

In conclusion, the software licensing of the Tesla Cybertruck is a masterclass in obfuscation, a tour de force of confusion. It's a document so impenetrable, so devoid of any semblance of clarity, that it could only have been created by a team of lawyers who were paid by the word. So, as you sit in your Cybertruck, remember: you may own the vehicle, but the soul of the machine belongs to Tesla. And they've got the software license agreement to prove it. Drive carefully, and maybe don't get too attached – you never know when your car might decide it's time for a change.

## Decoding the End User License Agreement, one headache at a time.

Embarking on the journey of understanding the End User License Agreement (EULA) of the Tesla Cybertruck is like deciding to climb Everest in flip-flops and a bathing suit – ill-advised, to say the least. The EULA is a document so long and convoluted, it makes "War and Peace" look like a light bedtime story. It's as if a group of lawyers were challenged to write a novel, and they decided the genre would be 'bureaucratic horror'.

Firstly, attempting to read the EULA is a test of human endurance. It's written in a font so small, you'd need a microscope to read it. It's the literary equivalent of trying to thread a needle while riding a rollercoaster – it's possible, but it's going to take a while, and you might throw up a bit.

As you delve into this behemoth of a document, you're greeted with language that suggests it was translated from English to Martian and then back to English by a computer with a vendetta against humanity. The legalese is so dense, each sentence feels like a riddle, wrapped in a mystery, inside an enigma, and then beaten to a pulp with a thesaurus.

The EULA covers everything – and I mean everything. From the software that makes the Cybertruck more than just a fancy garden ornament to the air you breathe while sitting inside it. It's like signing a contract with a genie – sure, you get your wishes, but the fine print might turn you into a toad.

And let's talk about updates. According to the EULA, Tesla can update the software in your Cybertruck whenever they want, for whatever reason they fancy. It's like living in a house where the landlord can come in and move the walls around every night. One day, you might wake up to find your Cybertruck now thinks it's a submarine or a spaceship – the EULA certainly doesn't rule it out.

Then there's the data collection clause. By agreeing to the EULA, you're essentially giving Tesla permission to know more about your driving habits than you do. They'll know where you've been, how fast you got there, and how many times you belted out 'Bohemian Rhapsody' at the top of your lungs. It's like having a stalker, but you're paying them for the privilege.

Attempting to understand the EULA is like trying to solve a Rubik's Cube in the dark. It's a maze of clauses, sub-clauses, and clauses about clauses. There are sections that contradict each other, sections that make no sense, and sections that you're pretty sure were included just for a laugh.

In conclusion, the EULA of the Tesla Cybertruck is not just a document; it's a journey into the heart of darkness, a labyrinth of legalese, a rollercoaster of bureaucratic gibberish. It's an encyclopedia of confusion, each page a fresh hell of complexity. So, when you sit down to try and decode this monolithic text, remember: it's not meant to be understood by mere mortals. It's there to remind you that, in the grand scheme of things, you know nothing, Jon Snow. And with that, I wish you good luck – you're certainly going to need it.

## Why your car's software thinks it's still living in the 1990s.

In the futuristic world of the Tesla Cybertruck, you would expect the software to be as cutting-edge as the vehicle's design, which looks like it was sketched by a child who's only ever drawn using a ruler. However, the reality is that the software seems to have been time-warped from the 1990s, a period when the height of technological sophistication was a Nokia 3310.

Let's begin with the infotainment system, a term I use as loosely as the term 'salad' at a fast-food restaurant. This system is less 'infotainment' and more 'misinformation'. It operates with the speed and agility of a tortoise wading through treacle. Trying to get it to respond to your commands is like trying to have a conversation with a moody teenager – you know there's intelligence there, but it's doing its best to hide it.

The navigation system is another marvel of ancient technology. It's reminiscent of a time when 'MapQuest' printouts were the height of navigational innovation. Inputting a destination into the Cybertruck's system and expecting a straightforward route is like asking a caveman for directions to the nearest Starbucks. You're better off navigating by the stars or following a trail of breadcrumbs.

Then there's the voice recognition feature – a feature so comically inept, it's like talking to a confused parrot. You ask it to find the nearest charging station, and it starts playing a podcast about the migratory patterns of the monarch butterfly. It's like those early voice recognition phone systems where you'd shout "Representative!" into the phone and end up being connected to a pizzeria in New Jersey.

Let's not forget the touchscreen response time, which has all the urgency of a sloth on holiday. You press something, go off to make a cup of tea, come back, and it might have registered your command. The lag is so profound; you wonder if the software is being powered by a hamster running in a wheel.

And software updates – oh, the updates. They arrive with the unpredictability of a British summer. When they do appear, downloading them is a day-long affair, reminiscent of trying to stream a video on dial-up internet. You half expect to hear the screeching tones of a modem when the download starts.

The pièce de résistance, however, is the Cybertruck's digital dashboard. It's like someone took the digital watches of the 90s and thought, "Yes, that's what we need for the dashboard." The pixelated display could be outclassed by a Game Boy. You almost expect to see Tetris blocks falling down as you drive.

In conclusion, the software constraints of the Tesla Cybertruck are a bewildering blend of nostalgia and frustration. It's as if the vehicle is caught in a time loop, where the body is speeding into the future, but the software is anchored firmly in the past. It's like having the sleekest, most modern TV available, but you can only watch reruns of 'Saved by the Bell'. So, as you sit in your Cybertruck, pressing buttons and waiting for something to happen, remember: patience is a virtue, and time travel is apparently possible – at least in the world of software.

## How to Lodge a Complaint Into the Void

When it comes to lodging a complaint about your Tesla Cybertruck, you must first understand that you're essentially shouting into the cosmic abyss. It's a bit like trying to send a strongly worded letter to a black hole and expecting a polite response. You, the proud owner of a vehicle that looks like it was designed by a polygon-obsessed lunatic, are about to embark on a journey of fruitless endeavors and Kafkaesque customer service.

The first step is, of course, finding someone to complain to. This in itself is as challenging as trying to nail jelly to the ceiling. You begin with a phone call, where you're greeted by an automated voice that offers you about as much empathy as a vending machine. The menu options are a labyrinthine puzzle, each choice leading you further down the rabbit hole of hold music and existential dread.

Once you finally navigate this maze and find a human being to talk to, you realize that this person is armed with a script so impenetrable, it makes the Enigma Code look like a children's crossword puzzle. They respond to your well-reasoned complaints with the enthusiasm of a sloth on sedatives. No matter what your issue is, their answers are as canned as spam, and about as satisfying.

But let's say you persevere and decide to take your complaint to the next level – the infamous email. Composing an email to lodge a complaint about your Cybertruck is like writing a Shakespearean sonnet that no one will ever read. You pour your heart and soul into detailing every issue, every nook and cranny of dissatisfaction. You hit send, and your email disappears into the ether, possibly to be read by a distant civilization in a thousand years, who will no doubt marvel at your quaint problems.

You might consider taking to social media, airing your grievances in the public sphere. This is the digital equivalent of standing on a street corner and yelling at clouds. Sure, a few passersby might nod in sympathy, but your words are just whispers in a hurricane of online noise. The Cybertruck, a vehicle that has more in common with an escaped piece of military hardware than a car, continues to baffle and bemuse in equal measure.

Now, let's not forget the possibility of a written letter, a method as archaic in this digital age as sending a carrier pigeon. You craft your letter, each word a testament to your frustration, and send it off. The letter probably ends up in a pile somewhere, serving as a makeshift coaster for a customer service rep's coffee mug.

In conclusion, lodging a complaint about your Tesla Cybertruck is an exercise in futility, a masterclass in the art of patience-testing. It's a journey into the heart of darkness, where hope goes to die and complaints are met with a shrug. But fear not, for you are not alone. You are part of a fellowship of Cybertruck owners, each as bewildered and bemused as you are. Together, you stand, shouting into the void, united in your bewilderment.

And who knows, maybe one day, someone will shout back. But I wouldn't bet my Cybertruck on it.

## The Art of Yelling at Clouds

Owning a Tesla Cybertruck is akin to being in a relationship with a brick wall that occasionally updates its own software. Voicing concerns about this vehicle is a practice in futility, an exercise that can be compared to yelling at clouds – therapeutic, perhaps, but ultimately as productive as trying to teach a goldfish to tap dance.

When you first voice your concerns, perhaps about the way the Cybertruck handles like a drunken elephant on roller skates, you'll be met with the kind of blank, soulless response usually reserved for politicians. It's as if your words enter a void, disappearing into the ether with as much impact as a marshmallow bullet.

Let's delve into the various methods you might employ in this noble yet fruitless endeavor. You might start with a phone call to customer service, where an automated system gives you more options than a gourmet restaurant menu. By the time you finally speak to a human, you've forgotten why you called and possibly your own name.

Then there's the option of sending an email, a process that feels like sending a love letter to a black hole. You carefully articulate your issues, pouring your heart and soul into every sentence, only for it to vanish into the digital abyss. The likelihood of receiving a response is about the same as finding a needle in a haystack, if the haystack were the size of Texas.

You might even attempt to use social media, broadcasting your grievances to the world. Here, your words will float in the vast ocean of online noise, lost among cat videos and people arguing about pineapple on pizza. It's a bit like shouting into a hurricane and expecting an echo.

And let's not forget the classic, the written letter. This method is as outdated as using a fax machine, but there's something charmingly quaint about it. It's like sending a message in a bottle – if the bottle were then lobbed into the Mariana Trench.

The art of yelling at clouds, or voicing concerns about your Cybertruck, is a masterclass in patience. It's a Zen exercise, teaching you the valuable lesson that sometimes in life, you just have to let go. It's about accepting that some battles cannot be won, that sometimes the best you can do is scream into the void and then go about your day.

In this process, you may find yourself experiencing a range of emotions. There's confusion – why is no one listening? There's frustration – why did I buy a vehicle that looks like it was designed by a polygon in a mid-life crisis? And finally, there's acceptance – the realization that sometimes, all you can do is laugh.

In conclusion, mastering the art of yelling at clouds when it comes to your Cybertruck is about finding peace in the chaos. It's about understanding that while you might not be able to change things, you can change the way you react to them. So go ahead, yell at those clouds, voice your concerns, and then take your Cybertruck for a spin – because, at the end of the day, it's still a marvel of engineering, even if it doesn't always listen.

## Crafting the Perfect Unread Email

Writing an email to the customer support of Tesla's Cybertruck is a bit like writing a love letter to a brick. No matter how heartfelt or desperate your pleas, they're about as likely to elicit a response as a microwave is to cook a three-course meal. But fear not, for there is an art to crafting these digital messages in bottles, an art that ensures they are perfectly tailored for their final resting place – the bottomless pit of the unread inbox.

Firstly, start with a catchy subject line. This is crucial. It needs to scream desperation and urgency, something along the lines of "Cybertruck Transformed into Decepticon" or "Help, My Cybertruck Has Kidnapped My Cat." This won't make any difference, of course, but at least you'll have the satisfaction of knowing you've been creative in your approach to being ignored.

Next, the greeting. It's important to strike the right balance between hope and despair. A simple "Dear Sir/Madam" is too mundane; you need something that grabs attention, like "Dear Overlord of Customer Service" or "To Whom It May Concern, If Anyone Actually Does." This sets the tone of respectful desperation that's crucial for an unread masterpiece.

Now, the body of the email. This is where you pour out your heart, detailing every fault, quirk, and unsolicited autopilot adventure your Cybertruck has taken you on. Remember, details are key. The more specific, the better. If your Cybertruck decided to play "Bohemian Rhapsody" at full volume at 3 AM, they need to know every lyric that was blasted through your neighborhood.

Be sure to include how the issue makes you feel. Phrases like "I weep daily in the driver's seat" or "I've started talking to the GPS for companionship" really paint a vivid picture of your suffering. Emotional blackmail is completely acceptable in this scenario; after all, you're crafting an email, not negotiating peace treaties.

Don't forget to add a touch of drama. Throw in some metaphors and similes. "Driving this Cybertruck is like steering a melancholic whale through a sea of molasses." It doesn't have to make sense; it just has to be memorable.

As you draw to a close, it's time for a call to action. This should be a blend of pleading and the faintest glimmer of hope that someone, somewhere, might read your words. Something along the lines of, "I await your reply like a parched desert awaits rain" should do the trick.

Finally, the sign-off. This is your last chance to make an impression, so make it count. "Yours despairingly" or "In eternal servitude" adds just the right amount of melodrama to your sign-off. It's the cherry on top of your email sundae of sorrow.

In conclusion, remember, the key to crafting the perfect unread email is to embrace the futility of the exercise. It's about finding joy in the little things — like the perfect metaphor or a particularly poignant turn of phrase. Your email may never be read, but in the process of writing it, you've embarked on a journey of catharsis, a therapeutic outlet for your Cybertruck-induced frustrations. And who knows, maybe, just maybe, your email will be the one that breaks through, the one that finds its way to a human eye. But probably not.

Printed in Great Britain
by Amazon